Melbourne's Water Catchments

PERSPECTIVES ON A WORLD-CLASS WATER SUPPLY

James I. Viggers, Haylee J. Weaver and David B. Lindenmayer

Dedication

This book is dedicated to my wife Diana Viggers for her unfailing support over many decades.

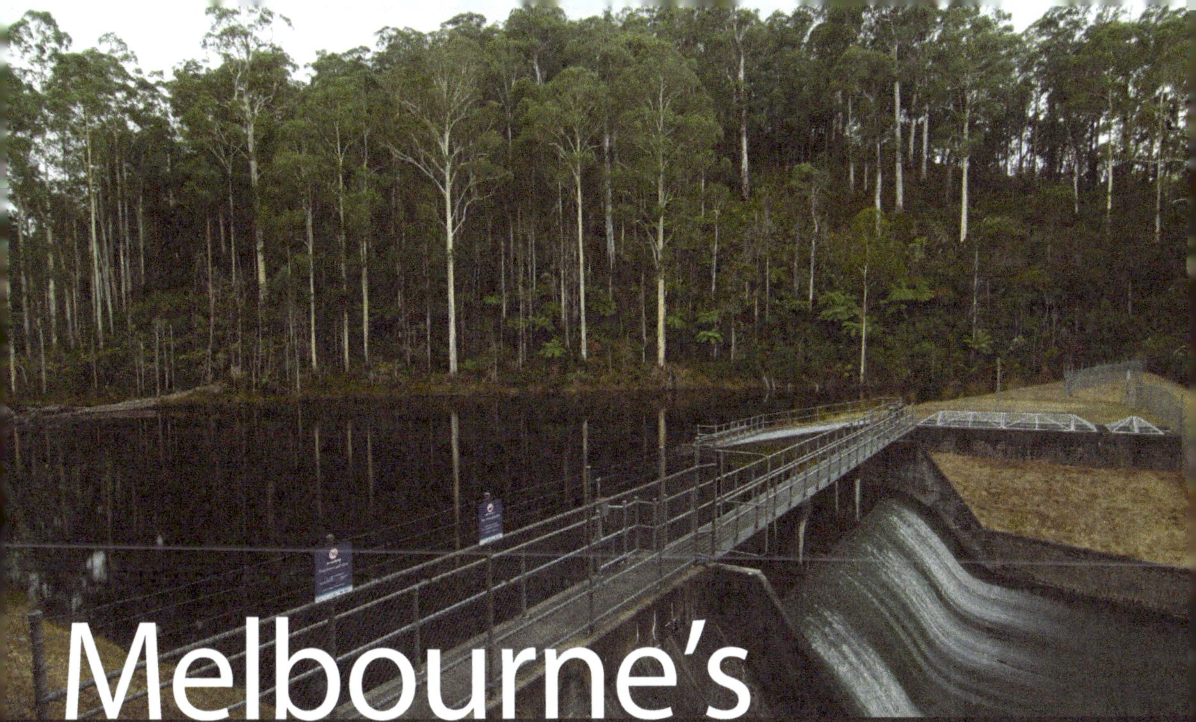

Melbourne's Water Catchments

PERSPECTIVES ON A WORLD-CLASS WATER SUPPLY

James I. Viggers, Haylee J. Weaver and David B. Lindenmayer

PUBLISHING

© James Viggers, Haylee Weaver and David Lindenmayer 2013

All rights reserved. Except under the conditions described in the *Australian Copyright Act 1968* and subsequent amendments, no part of this publication may be reproduced, stored in a retrieval system or transmitted in any form or by any means, electronic, mechanical, photocopying, recording, duplicating or otherwise, without the prior permission of the copyright owner. Contact **CSIRO** PUBLISHING for all permission requests.

National Library of Australia Cataloguing-in-Publication entry

Viggers, James I., author.

Melbourne's water catchments : perspectives on a world-class water supply / By James I Viggers, Haylee J Weaver and David B Lindenmayer.

9781486300068 (paperback)
9781486300075 (epdf)
9781486300082 (epub)

Includes bibliographical references and index.

Watershed management – Victoria – Melbourne.
Watershed management – Victoria – Melbourne – History
Water quality management – Victoria – Melbourne.
Melbourne (Vic.) – Water supply.

Weaver, Haylee J., author.
Lindenmayer, David, author.

333.910099451

Published by
CSIRO PUBLISHING
36 Gardiner Road, Clayton VIC 3168
Private Bag 10, Clayton South VIC 3169
Australia

Telephone: [+613] 9545 8555
Local call: 1300 788 000 (Australia only)
Fax: +61 3 9662 7555
Email: csiropublishing@csiro.au
Website: www.publishing.csiro.au

Front cover: (top) Armstrongs Creek (D. Blair); (bottom) Network of water supply reservoirs for Melbourne (C. Hilliker)

Set in 12/14 Minion
Edited by Adrienne de Kretser, Righting Writing
Cover design by Andrew Weatherill
Text design by James Kelly
Typeset by Thomson Digital
Index by Indexicana
Printed by Ingram Lightning Source

Feb26_RP_ILS

CSIRO PUBLISHING publishes and distributes scientific, technical and health science books and journals from Australia to a worldwide audience and conducts these activities autonomously from the research activities of the Commonwealth Scientific and Industrial Research Organisation (CSIRO). The views expressed in this publication are those of the author(s) and do not necessarily represent those of, and should not be attributed to, the publisher or CSIRO. The copyright owner shall not be liable for technical or other errors or omissions contained herein. The reader/user accepts all risks and responsibility for losses, damages, costs and other consequences resulting directly or indirectly from using this information.

Contents

Preface

The city of Melbourne – Australia's second-largest city – has some of the best-quality drinking water in the world, but few Melburnians have any idea where their water comes from or how the water supply and sewerage system was developed. The world-class system that services Melbourne did not come about by good luck and chance. Rather, it is the product of careful planning, outstanding engineering and construction, and extraordinarily far-sighted policies which were decades, if not centuries, ahead of their time, such as the closure of catchments to most human activities.

This book documents the history of the development of the water supply for the City of Melbourne. The first three chapters, which comprise Part 1, offer a brief historical account from the arrival of the first settlers to the early construction of large dams, and sketch the problems arising from disease, pollution, repeated droughts and bushfires. These chapters also tell of the lives of the visionary engineers and planners who tackled these issues to secure the city's water supply through the design and construction of numerous weirs, dams and extensive networks of pipelines and aqueducts. They also encompass the lives of the more humble but nevertheless important people who were involved in maintaining the system, such as those who manned fire towers and serviced aqueducts.

The second part of the book (Chapters 4–6) are a 'water man's' perspective on the continued expansion and further development of Melbourne's water supply from the 1960s until the present day. Jim Viggers (Figure A) was an Operations and Maintenance Engineer and a Senior Executive for Melbourne's water supply for 20 years, and an employee of the Melbourne and Metropolitan Board of Works for over 35 years. This book provides a series of unique and fascinating insights into a critical part of Australia's natural resource management infrastructure.

Figure A: Jim Viggers (image: Shire of Healesville).

The final chapter offers perspectives on the many vexed issues surrounding the construction of the controversial North-South Pipeline and the desalination plant, including the long-term assessments made of these projects by water authorities many years before their commencement. This chapter also considers how and why past approaches to water infrastructure management differ from current practices.

Our hope is that this book will provide a new set of insights on why and how Melbourne established such a world-class water supply. We also hope it will stimulate discussion on future directions in light of the major challenges that will arise in the coming 20–50 years.

Jim Viggers
Haylee Weaver
David Lindenmayer
March 2013

Acknowledgements

Many people provided important information that enabled this book to be completed. These include employees of the former Melbourne and Metropolitan Board of Works (the MMBW) or its reincarnation (Melbourne Water), including Frank Barnes, Paul Balassone, Austin Byrne, Rob Cranston, Steven Ennor, Janet Irving, Graeme Jackson, Roy Kermode, John Langford, Frank Lawless, Russell Read, Ian Smith and Neville Smith. Additional key information was provided by Jason Odering (Royal Historical Society of Victoria).

The Honourable Simon Crean MP and his staff provided support in locating key historical documents.

Claire Shepherd and Clive Hilliker (at the Australian National University) assisted with editing and graphic design. Karen Viggers provided expert editorial insights.

David Blair (ANU) kindly dedicated several days to taking high-quality photographs.

John Manger from CSIRO Publishing championed this project.

Glossary

Adit The entrance to a tunnel (a term often used in mining).

Alienated land Areas that are used for purposes other than, or in addition to, water supply (e.g. logging). Closed catchments were typically unalienated land.

Aqueduct An open-air channel to convey water.

Catchment An area of land that water falls on, to feed rivers or other watercourses.

Confluence The meeting point of two watercourses.

Dam A wall built to halt the flow of water down a river. Also see **Embankment**.

Diversion A barrier placed on a river to move water in a new (artificial) direction, i.e. into an aqueduct or pipeline (similar to a weir).

Draft (also called draw-down) The amount of water being removed from a reservoir by consumers.

Embankment A wall built to contain water. Can be across a river (a dam), a valley or other suitable area.

Evapotranspiration The process by which trees extract water from soils and the subsequent loss of the water from the system as the trees transpire.

Gravity-fed system Water moving from a source at an elevation higher than the destination of that water.

Headworks Term used to describe catchments and storage reservoirs, i.e. where water is harvested and stored at the uppermost part of a water supply chain.

Hydrology The study of the movement of water.

Main (usually plural, mains) The system of large pipes to convey water, usually underground.

Off-stream A water storage facility that was not the result of damming a river or watercourse. Often does not have a catchment, i.e. water is piped into it from another source.

Reservoir An artificial lake created by damming a watercourse or building an embankment.

Reticulated water The supply of water through pipes from a water utility or business.

Siphon (often called inverted siphon) A bridging mechanism to allow aqueducts to span valleys or other structures while following the same contour.

Streamflow The flow of water in streams, rivers and other channels; this is a major element of the water cycle.

Transpiration The loss of water vapour from plants as they respire.

Watershed see **Catchment**.

Weir A barrier placed across a watercourse to alter its flow. Also see **Diversion**.

Yarra tributaries The collective name given to Starvation, McMahons and Armstrongs creeks.

Conversion table

Metric units have been used throughout the book, with Imperial conversions where historically appropriate. For context, a standard Olympic-sized swimming pool holds ~2.5 ML, and Port Phillip Bay holds ~2500 GL.

1 litre (L) = 1000 millilitres (mL)
1 kilolitre (KL) = 1000 L
1 megalitre (ML) = 1000 KL (1 million L)
1 gigalitre (GL) = 1000 ML (1000 million L)

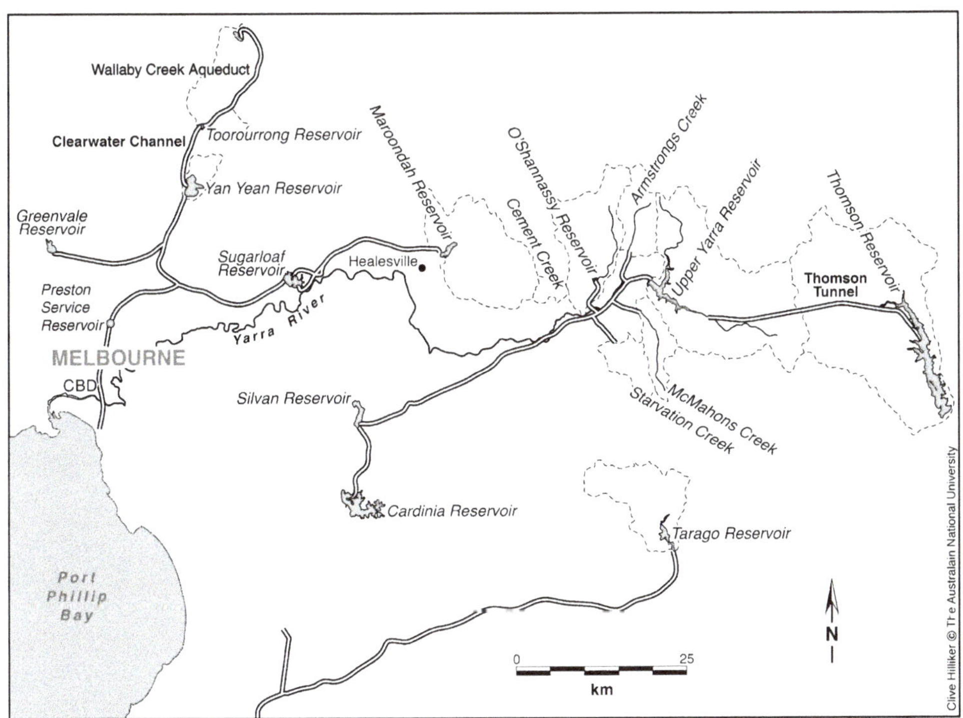

Figure B: The network of water supply reservoirs for Melbourne (image: C. Hilliker).

Part 1
EARLY HISTORY: PRE-1900 TO 1960

Chapter 1: Pre-1900

Introduction

Today, almost all of Melbourne's drinking water comes from reservoirs located in closed catchments. Human access is largely excluded from these areas to protect the city's water supply. Such a system of closed catchments has almost no parallel, either nationally or globally. To understand how Melbourne's water supply system has developed, in this first chapter we go back to the beginning of the settlement of the city and explore some of the key reasons water quality was such an important issue.

Water for a new settlement

Water was first supplied to the fledgling settlement of Melbourne from the Yarra River. The settlers that arrived in 1835 positioned their campsites on the river's banks. A natural rock ledge across the river, near the present-day Queens Bridge, separated fresh water upstream from the tidal sea water of Port Phillip Bay. This ledge, called 'The Falls' by locals (Figure 1.1), did not always exclude sea water at very high tides. Therefore, sites for drawing water were placed further upstream.[1] Enterprising people drew water from the Yarra River and sold it in barrels to other settlers.

As Melbourne expanded, the amount of pollution being washed into the river increased and the quality of the water deteriorated. There was no sewerage or drainage system for the growing township. Human waste and other pollution flowed freely down the streets, accumulated in low-lying areas and made its way into the Yarra River. Outbreaks of infectious diseases such as typhoid were common.

The Melbourne City Council, formed in 1842, was concerned about how clean water could be supplied to the rapidly expanding city.[2] Around this time in England, a revolution in water supply was occurring. Fresh water was piped to central points in streets and separated from disease-causing sewage and pollutants.[3] The Council sought similar kinds of infrastructure

Figure 1.1: An 1838 view of The Falls on the Yarra River, at Queen St. The Falls were physically removed from the Yarra River in 1883 (image: Public Record Office Victoria).

for Melbourne, and began evaluating proposals for the design of such a scheme. Early proposals involved the development of a steam-driven pumping station at Dights Falls on the Yarra River upstream of Melbourne (nowadays part of the Yarra Bend Park in Kew), with water to be piped to large tanks at Eastern Hill and Flagstaff Gardens in the city centre for distribution to consumers (Figure 1.2).[3] A water supply tank constructed at Eastern Hill in 1854 used a stream-driven pump to draw water from the Yarra River at the end of Spring Street.

Blackburn's water supply ideas

The first major transformation of Melbourne's water supply was shaped by ex-convict James Blackburn (Box 1.1), who arrived in Melbourne from Tasmania in 1849. In response to the state of the water supply to the city – which by then had a population of over 20 000 – Blackburn formed a company that was given a permit by the Melbourne City Council to install a pumping station upstream of The Falls (the rock ledge) on the Yarra River. Blackburn's company used steam-driven pumps to divert water from the

Figure 1.2: Dights Falls on the Yarra River. Dights Falls was the initial site of a proposed water supply scheme for Melbourne (image: Public Record Office Victoria).

river into elevated tanks and a charcoal purification plant on the corner of Flinders and Elizabeth streets.[4] The company sold clean water to carters at a rate of 1 penny per load, as per the permit's stipulations. Up to 700–800 cart loads of water could be sold per day in peak periods during summer.[4,5]

Following Blackburn's success with the project on the Yarra River, the Melbourne City Council appointed him City Surveyor and he was given the task of designing a permanent water supply scheme for the city.[3] Blackburn identified flaws in the Dights Falls proposal – its susceptibility to floods, the unreliability of the machinery needed to pump the water, and the limited area it could supply.[3] He set about investigating alternatives and completed

Box 1.1: James Blackburn: the father of Melbourne's water

James Blackburn (1803–1854), an English surveyor, architect and civil engineer, was transported for life to Australia for forging a cheque in 1833. He arrived in Hobart in late 1833 and his family joined him in 1835.[18] Blackburn worked for the Department of Roads and Bridges in Tasmania from 1833 to 1839 and was responsible for many surveying and road construction projects around Hobart and throughout Tasmania. He was granted a free pardon in 1841, and established an architectural design company in Hobart. Blackburn designed many buildings including churches and government public offices in Hobart, as well as the water supply system for Launceston, before moving to Melbourne in 1849.[18] Blackburn worked as a consulting engineer to Matthew Jackson on the Yan Yean Reservoir construction project. It was speculated that he had been given the lesser position because he was unable to take a full workload due to injuries sustained after falling from his horse in 1852,[19] perhaps while riding the city limits in an organised ride called 'beating the bounds'.[20] Blackburn did not live to see the completion of his vision for water supply. He died from typhoid in Melbourne in 1854.[18]

James Blackburn (image: Melbourne Water).

surveys around the Melbourne region, including at Mt Macedon and Lilydale. In 1850, he recommended that water from the Plenty River be used as the source of water for Melbourne as there were many streams and creeks of clean water that flowed into the Plenty River. In a more detailed report in 1851, he recommended Ryders Swamp (also called Riders Swamp) at Yan Yean north of Melbourne as the site for an off-stream storage to hold water from the Plenty River.[1] Blackburn also suggested that an open-air aqueduct could bring water from the Plenty River to Melbourne, with a reservoir near Pentridge Prison.[3] However, a reservoir was never built at that site. The Plenty River was favoured because of its proximity to Melbourne and its elevation meant that the water could be gravity-fed to the city (see Box 1.3). Blackburn envisaged water being piped directly to houses rather than to communal taps in central areas, which was the model in parts of Europe at the time.[3]

The colony of Victoria was established in 1851, with Melbourne as its capital. The colonial government of the day decided that the development of water and sewerage systems needed to be managed by a body other than the City Council. On this basis, a 'select committee on the sewerage of and supply of water for Melbourne' was appointed.[3] The committee initially preferred the Dights Falls plan, but became convinced of the need for the

1842: Melbourne Town Council formed.
Responsible for water supply for the city.

1853: Board of Commission of Sewers and Water Supply.
Formed by Legislative Council (following formation of the colony of Victoria).

1859: Board of Land and Works formed to include Commission of Sewers and Water Supply.

1862: Water Supply Branch of Department of Public Works took over water supply after Board of Land and Works was abolished.

1872: Water Supply Board formed.
Answerable to state government, tasked with developing water supply.

1891: Melbourne and Metropolitan Board of Works (MMBW, or 'The Board') formed.
Responsible for water supply, sewerage and drainage.

1992: Melbourne Water formed. Result of a merger between MMBW and other water suppliers for outer Melbourne.

Figure 1.3: Water supply was managed by a series of different agencies, changing frequently until the development of the Melbourne and Metropolitan Board of Works (MMBW), which remained for 100 years (image: C. Hilliker).

reservoir at Yan Yean after engineers advising the committee inspected both sites. The engineers supported Blackburn's plan and one committee member, Charles Oldham, augmented the proposal by suggesting that all (instead of only some) of the water in the Plenty River be diverted into the storage on Ryders Swamp, and that the water be conveyed to Melbourne in pipes rather than an aqueduct.[3] Despite resistance and concerns about whether enough labour could be sourced in the middle of the gold rush gripping Victoria at the time,[3] the committee supported Oldham's recommendations and suggested changes to Blackburn's plan, which was then approved for implementation.

The Select Committee also recommended that a specific body of commissioners be appointed to manage the water supply and sewerage systems. This became the Board of Commission of Sewers and Water Supply in 1853, and one of its first tasks was to develop the Yan Yean scheme.[2] The body in charge of water supply changed several times over the following 50 years (Figure 1.3).

The Yan Yean scheme

Construction of the Yan Yean scheme commenced in mid-1853. Matthew Bullock Jackson (Box 1.2), a young English engineer recently arrived in

Melbourne, was awarded the position of Resident Engineer for the project.[1] Jackson was responsible for the design and construction of Blackburn's proposal to harvest water from the Plenty River. Despite having proposed the initial idea, Blackburn was not given the Resident Engineer job but was – at Jackson's request – instead appointed as a consulting engineer.[1,3] From Jackson's own account, he and Blackburn 'worked together cordially' on the project.[6]

Blackburn's original proposal for Melbourne's water supply had been based on an estimated 182 L per person per day for the 70 000 inhabitants for Melbourne.[6] As the population of the city was expanding rapidly, Jackson increased the capacity of the scheme to supply 200 000 residents at 136 L per person per day.[3] Jackson also elected to draw water from the Plenty River to the reservoir downstream of the nearby swamps and the township of Whittlesea. This was cheaper than doing so upstream of the swamps, but it yielded discoloured and polluted water.[3]

The Yan Yean Reservoir was created by constructing a large earthen embankment, with an impervious puddle-clay core. The impervious core was the keystone of the dam's integrity. Local clay was excavated and transported to the dam wall to form the core. There, the clay was spread by hand into a thin layer, watered for 12 hours then compacted under the feet of workmen who trampled it for hours.[3] The laying and trampling process was repeated until the core wall reached its designated height. The core wall was very thick to withstand the pressure of holding the water in the reservoir. Work was done by hand and it was made more difficult because of the scarcity of labour caused by the gold rush at the time. Costs blew out as workers had to be paid more to work on the dam site than they would make at the goldfields.[3] While the embankment and pipe-laying works were underway, the water pipes forming the reticulation supply were being laid from central mains in city streets to individual houses.[1] During the construction of Yan Yean Reservoir, a temporary supply of water to Melbourne residents was derived from a tank erected on Eastern Hill in 1854 that was fed by pump from the Yarra River.[7]

The pipeline from Yan Yean to Melbourne was constructed in three sections. The pipes used were cast iron and lead-jointed, and were imported from England. The first section, from Yan Yean to Morang, comprised 762 mm diameter pipes, while the second section, from Morang to Preston, was 685 mm in diameter. The third section, from Preston to Melbourne, was 609 mm in diameter. Pressure-reducing valves were installed at each change of pipe size. These valves were designed to reduce in stages the pressure that built up in the pipeline as the water travelled from Yan Yean at 182 m above sea level, to the city at sea level.[3] The pressure-reducing valves did not perform well, and Jackson insisted that valves located in the lower part of Melbourne always be left partially open to prevent pipes from bursting due

Box 1.2: Matthew Bullock Jackson, engineer of Yan Yean Reservoir

Matthew Bullock Jackson was 27 when he arrived in Melbourne from England in 1852. He had come to Australia to oversee the construction of a railway line from Adelaide to Port Adelaide, but by the time he arrived the work had been postponed. This was due to the discovery of gold in parts of Australia, which led to the majority of the potential workforce leaving for the goldfields.[3] Little is known of Jackson's life before he came to Australia, except that he worked as a surveyor and draughtsman in England, including a role in developing railway infrastructure for the Lighthouse Stevensons.[3] He had also reported on the failure of an earthfill dam in Yorkshire, which may have been one reason he was appointed to manage the design and the construction of the new Yan Yean earthfill dam and associated downstream pipelines.[3]

Jackson's construction of Yan Yean Reservoir to cope with Melbourne's burgeoning population was tarnished somewhat by events following its completion. The water mains and domestic pipes kept bursting – an outcome of ineffective pressure-regulating valves installed in the large pipes. Most of the catchment from which the water was sourced was utilised for agriculture and other activities causing pollution. Water quality was poor, and typhoid was still prevalent in Melbourne even though water was not sourced directly from the polluted Yarra River. Faulty lining of the lead pipes used between street mains and houses also allowed lead to leach into the water and caused lead poisoning.[5] Jackson suffered from allegations of poor bookkeeping, as the cost of the project, predicted to be £438 178, blew out to £754 206.[21] Questions were asked about cost overruns. The public, spurred on by articles published in *The Argus*, were convinced that money had been spent on unnecessary extravagances. This was despite Jackson arguing that several costs had not been included in the original estimate of the project, including management and salary costs, legal expenses, acquisition of land and 100 km of pipes for the reticulation supply.[3,21]

Jackson returned to England in 1861, after several inquiries into the water supply for Melbourne were highly critical of decisions he made during the construction of Yan Yean Reservoir and subsequent water supply and quality problems. Before he left, and under duress, he wrote a letter of apology to an aggrieved contractor. He complied because he had been threatened with not being allowed to leave the colony. Jackson later wrote that he regretted writing the apology.[21] After returning to England, Jackson gave evidence following the failure of a dam in Sheffield and designed an alternative water supply scheme for the area.[21] Despite the humiliating circumstances of his time in Melbourne following the construction of Yan Yean, Jackson's legacy lived on – Yan Yean Reservoir still exists today (Figure 1.4).

Figure 1.4: Yan Yean Reservoir panorama (image: MMBW[8]).

to the build-up of pressure.[3] One of these pressure-reducing valves still exists today at Preston Reservoir, but it is inoperative.

The pipeline initially ran unbroken from Yan Yean Reservoir to Melbourne. However, as foreshadowed by Jackson during the development of Yan Yean, a service reservoir was constructed along the pipeline in 1864.[3] The Preston Reservoir, as it was called, was built in response to poor water pressure in Melbourne during peak water demand periods (Figure 1.6).

Box 1.3: An underlying principle of a gravity-fed water supply – what you see is not really what you get

Much of this book focuses on the construction of the various components of infrastructure that comprise the gravity-fed water supply system for the Melbourne and metropolitan area. Key to any gravity-fed water supply system is a series of large reservoirs at elevations above sea level (Figures 1.5, 1.6). The weight of the water in a pondage creates pressure which pushes the water to other reservoirs and then on to consumers. A critical principle in a gravity-fed system is a minimum operating level (MOL). This is the lowest level to which water in a reservoir can fall before difficulties in water supply begin to occur. Taking too much water from a reservoir will reduce the water pressure in the system, which will affect supply. In the early 1900s there was difficulty in maintaining the supply of water to residents in the developing hilly eastern suburbs of Melbourne due to insufficient pressure (see Chapter 2). This would also happen today if storage reservoirs were drawn below their MOL. Silvan Reservoir, for example, has an MOL of 32.61 GL of its capacity of 40.45 GL (see Table 1.1). This means that 80% of the water in Silvan Reservoir has to remain in it to maintain the pressure. Conversely, Maroondah Reservoir's MOL of 2.2 GL represents only 9% of its capacity, meaning that the majority of the water in Maroondah Reservoir can be used for supply and only a small amount must remain in the reservoir to maintain pressure. It is important to understand MOLs, as bald statements in the media such as 'Melbourne's dams are 39% full' do not reflect how much water is actually available for use.

Table 1.1. Some of Melbourne's water storages' capacity and MOL

Each water storage has a different MOL because of differences in topography, underlying geology, water quality, downstream supply pressure and other factors.

Storage	Capacity (GL)	MOL (GL)	MOL as a % of total supply
Thomson	1068.00	158.87	15
Cardinia	286.91	139.22	48
Upper Yarra	200.58	70.70	35
Sugarloaf	96.25	21.50	22
Silvan	40.45	32.61	80
Tarago	37.58	4.00	11
Yan Yean	30.26	6.00	20
Greenvale	26.84	15.70	53
Maroondah	22.18	2.20	9
O'Shannassy	3.12	1.52	33
TOTAL CAPACITY	**1812.17**		

Water mains frequently burst due to increasing pressure from the gravity-fed system resulting from the malfunction of Jackson's pressure-reducing valves. Taps would run dry during peak times, and there was excess water available during off-peak periods. Water that flowed into Preston Reservoir during off-peak periods could be stored there for use during the peak times. This maintained continuity of pressure and supply to consumers across the city and the frequency of burst mains reduced significantly.

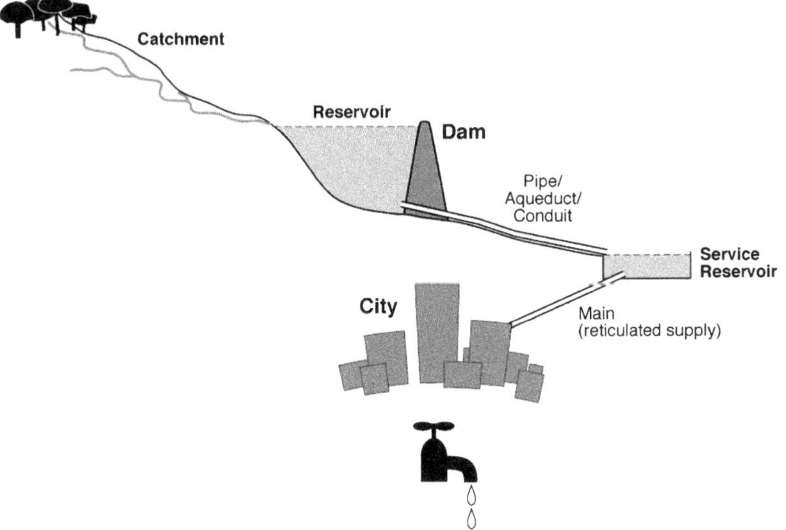

Figure 1.5: A schematic diagram showing the progression of water from catchments to households (image: C. Hilliker).

Figure 1.6: Cleaning the Preston No. 1 Reservoir. Caretaker Wally Cumberland Sr supervises workers removing silt from the bottom of the reservoir (image: MMBW photo by George Self).

The pipeline between Preston Reservoir and Collingwood was duplicated in 1869 to increase the supply of water to the growing city.[1] More water was thus needed from Preston Reservoir, so the 762 mm pipeline from Yan Yean Reservoir to Morang was lifted and replaced with an aqueduct that fed into a new pipe-head reservoir at Morang. The lifted pipes were then re-laid between Morang and Preston Reservoir, thereby duplicating the existing 685 mm pipeline between those locations.[1]

Closure of the Yan Yean water catchment

The construction of Yan Yean Reservoir gave Melbourne a reliable supply of drinking water. However, there was no sewerage or drainage system, and the health of the city's residents was still at risk from infectious diseases.

In 1860, a parliamentary inquiry into water quality at Yan Yean found that Bruces Creek, a tributary of the Plenty River running into Yan Yean Reservoir, had been badly polluted by human activities in Whittlesea, but this was ignored by water planners and politicians until the 1880s.[3] Until then, contaminated water continued to flow into Yan Yean Reservoir. At that time, the link between activity in catchments and water quality was beginning to be understood and the concept that the forests themselves contributed to improved water quality was very new. The idea of preserving forests to 'encourage rainfall' had been suggested by George Perkins Marsh in his book *Man and Nature* (1864) and the men of the Water Supply Board enthusiastically accepted the idea that there was a direct relationship between tree cover and rainfall.[3,9] The sawmills and log-splitters working in the Yan Yean area had consumed large areas of forest. The drought of 1865–66 also reduced runoff into the Plenty River and the water level in the reservoir fell, which concerned the Board.[10] It was essential that as much water as possible flow into the reservoir.

In 1872, in light of increasing evidence that Yan Yean water was contaminated before its entry into the reservoir, together with continued logging in the catchment reducing tree cover, the Water Supply Board declared that parts of the Yan Yean Catchment should be permanently reserved for water supply and that all timber cutting was banned within those areas.[3]

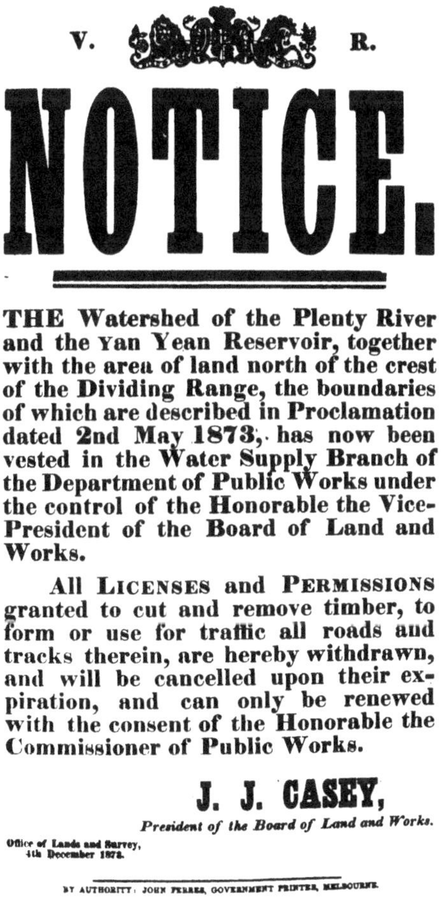

V. R.

NOTICE.

THE Watershed of the Plenty River and the Yan Yean Reservoir, together with the area of land north of the crest of the Dividing Range, the boundaries of which are described in Proclamation dated 2nd May 1873, has now been vested in the Water Supply Branch of the Department of Public Works under the control of the Honorable the Vice-President of the Board of Land and Works.

All LICENSES and PERMISSIONS granted to cut and remove timber, to form or use for traffic all roads and tracks therein, are hereby withdrawn, and will be cancelled upon their expiration, and can only be renewed with the consent of the Honorable the Commissioner of Public Works.

J. J. CASEY,
President of the Board of Land and Works.

Office of Lands and Survey,
4th December 1873.

BY AUTHORITY: JOHN FERRES, GOVERNMENT PRINTER, MELBOURNE.

Figure 1.7: Sign advising cessation of logging within the Yan Yean Catchment, 1873 (image: Melbourne Water).

This decision was not popular among timber workers or others in the Whittlesea area who worked in and around the catchment.[3]

The water catchment areas supplying Yan Yean Reservoir were closed progressively from the early 1870s, paving the way for future water catchments to be reserved for water supply purposes from their inception (Figure 1.7). Controversies surrounding closure of catchments would continue each time it was proposed that land be reserved for water supply purposes. Parts of the southern areas of the Great Dividing Range, within closed water supply catchments, now support some of the most pristine forests in Australia.[11]

In 1880, a committee investigating the quality of water from Yan Yean heard evidence from William Davidson (Box 1.4), Superintending Engineer of Water Supply for the Department of Public Works, on the pollution of the water supply to Yan Yean from farms and land use in the Whittlesea area.[9] This committee recommended to the state government that new diversions be constructed, such as an aqueduct from Wallaby Creek to Yan Yean, to bypass the polluted areas of Whittlesea that were contaminating the water supply.

Challenges to maintaining adequate water supply

Although Yan Yean was one of the largest reservoirs in Australia at the time, the population of Melbourne was increasing so quickly that consumer demand began to outstrip the supply of water. By 1875, the population exceeded 245 000 – far more than the 200 000 people that Jackson had designed the system to supply.[7] The Water Supply Board examined alternative sources of supply and identified that the Watts River to the east of Melbourne would be able to provide ample, clean water.[3] The Watts River Catchment was predominantly unalienated forest. Use of land for water harvesting was controversial at Yan Yean, because the closure of the catchment forced out timber cutters. It was recommended to the government that the Watts River Catchment be closed to circumvent arguments over land use.[9]

Despite the work on assessing the suitability of the Watts River Catchment, no action was taken in subsequent years to include it as part of Melbourne's water supply infrastructure. However, this changed following a large flood in March 1878 which washed away the bridge supporting the Yan Yean Aqueduct across the Plenty River at Mernda, cutting the supply of water to Melbourne. William Davidson immediately implemented work to build a

Box 1.4: William Davidson, water catchment crusader

William Davidson (1844–1920) was born in Ireland. He left school at 13, worked at various jobs and in 1859 travelled from Liverpool to Melbourne. Following his arrival, Davidson walked to Ballarat and commenced work with an uncle who was a surveyor.[12] Davidson was perhaps the most important person in the history of the development of Melbourne's water supply system. He worked to expand the water supply to meet the demands of a growing population by commissioning new catchments. Davidson developed what was to become known as the closed catchment policy (Box 1.5), where water catchments were closed to almost all human activity and therefore dedicated to the sole purpose of harvesting water. Together with William Thwaites (Box 1.6), Davidson secured a clean water supply for Melbourne through modifications and expansion of the Yan Yean scheme and closure of its catchment. Davidson was instrumental in achieving the reservation (and later vestment) of the Maroondah, O'Shannassy and Upper Yarra catchment areas for the sole purpose of water supply. This was particularly noteworthy as it was not without fierce opposition from the timber industry. In 1891, Davidson was appointed to the position of Inspector-General of Public Works and Chief Engineer of Water Supply of the MMBW.[12] Davidson's influence in water supply decision-making was still evident in 1909, when he was invited to serve on a board to make recommendations about the adoption of the Maroondah and O'Shannassy schemes.[9,12] Davidson retired from the MMBW in 1912 and in later life reflected on his unorthodox path to the offices he held, as there was a shift towards knowledge and qualifications obtained from university study rather than on-the-job experience.[12] Davidson died in Melbourne in 1920.

William Davidson (image: Public Record Office Victoria).

wooden flume to replace the lost section of the aqueduct and restore supply of water to the city.[3]

Davidson supervised the construction team, which worked non-stop for three days and nights until the flume was replaced and the water supply to Melbourne was restored.[12] While the supply from Yan Yean Reservoir was disrupted, the old tank at Eastern Hill in central Melbourne was recommissioned with water pumped from the Yarra River, thereby providing a limited supply to the city.[3] Following completion of the repairs and the resumption of the water supply to the city, Davidson was hailed as a hero.[12]

The failure of the single supply of water placed increased importance on finding alternative locations and sources of water for Melbourne.

Figure 1.8: a) The Clearwater Channel conveyed water from Toorourrong Reservoir to Yan Yean Reservoir, bypassing the polluted Plenty River (image: MMBW [8]). b) The Clearwater Channel, present day (image: D. Blair).

Box 1.5: Closed catchments – decades ahead of their time

The visionaries for Melbourne's water supply who made the tough decisions to close areas of forest to maintain the integrity of the catchments were many decades ahead of the rest of the world. Indeed, it was not until the 1990s that water planners for cities like New York fully appreciated the ecological and economic value of protecting water catchments. For example, in New York, the cost of buying land to protect the water catchments was found to be significantly less than the costs of building and maintaining water treatment plants. Some have suggested that investment in preserving water catchments is one-tenth the cost of fully treating water (e.g. http://www.ecosystemmarketplace.com/pages/dynamic/article).[22] For instance, the cost of treating water from a catchment with 60% forest cover is half the cost of treating water from a catchment with 30% cover and one-third of that with 10% cover.[22] Notably, some of Melbourne's water catchments have excluded human access for a prolonged period, meaning the significant expense of buying private (alienated) land to secure catchment integrity is not required.

The Watts River proposal was revisited, with a 1879 survey of the area to find a site for the diversion weir and an aqueduct.[1] In 1880, the Water Supply Board presented a report to the government recommending a diversion on the Watts River, with the additional recommendation of a dam on the river at a later time as a substantial source of water.[1]

Construction of new water supplies to Yan Yean Reservoir

In response to the state government's report on securing an unpolluted source of water for Yan Yean, in 1880 Davidson dispatched a young engineer, William Thwaites, to investigate the potential of harvesting water from King Parrot Creek, a tributary of the Goulburn River.[3,13] Thwaites decided that Wallaby Creek (a tributary of King Parrot Creek) was a better option as it was higher in elevation. He recommended to Davidson that the best action would be a diversion weir on Wallaby Creek with an aqueduct to a saddle of the Great Dividing Range to send the water to Yan Yean Reservoir.

Work on the aqueduct commenced under Thwaites' supervision in 1883.[13] Thwaites also designed and supervised the construction of Toorourrong Reservoir on East Plenty River, a holding facility for water between Wallaby Creek and Yan Yean Reservoir, and the Clearwater Channel, which connected Toorourrong to Yan Yean Reservoir via the inlet channel and tunnel constructed by Jackson in 1853–57 (Figure 1.8).[13] In 1886 the Wallaby Creek Aqueduct was extended to capture water from Silver Creek, another tributary of King Parrot Creek. The Clearwater Channel bypassed the township of Whittlesea where the Plenty River became polluted, and became

Box 1.6: William Thwaites, a saviour of Melbourne

William Thwaites (1853–1907) was born in Melbourne in the same year that construction began on the Yan Yean Reservoir. He earned an engineering degree at the University of Melbourne, and later served on the University's Board. Thwaites was intelligent and an exceptionally hard worker and his designs functioned well into the 20th century. For example, he designed the pumping station that feeds water to the Botanic Gardens from Dights Falls on the Yarra River.[13,23] In 1883, Thwaites was appointed Engineer of Roads, Bridges and Reclamation Works, enabling him to become more involved with the development of Melbourne's sewerage and drainage systems.[24] He was the first Engineer-in-Chief of the MMBW following its inception in 1891, and oversaw the development and implementation of the sewerage system that he had designed.[23] Thwaites died unexpectedly in 1907 from pneumonia. His lasting achievement was the supply of clean water and sanitation to Melbourne through his careful design and construction of new water supplies and development of the sewer system.

William Thwaites (image: Melbourne Water).

the source of supply of water to Yan Yean Reservoir.[2] An important feature of Wallaby and Silver creeks was that they were located in uninhabited areas. With the new aqueducts and the channel from Toorourrong Reservoir to Yan Yean, water quality was greatly improved.

Around the time that the aqueducts and the Clearwater Channel were being built, Thwaites was organising the draining of swamps, such as at Elwood, and other wet areas around metropolitan Melbourne where pollution and waste pooled.[5] He played a key role in developing the city's sewerage system, in the 1890s.

Diversion of the Watts River and construction of the Maroondah Aqueduct

In 1876, the Water Supply Board recommended that the Watts River Catchment be permanently reserved for the supply of water.[9] Davidson designed an aqueduct to take water from a diversion weir on the Watts River at Healesville to the Preston Reservoir in Melbourne. The design also incorporated mechanisms to accommodate the increased flows that would be needed in future. This foresight paid off when the Maroondah Reservoir was constructed in 1927 (see Chapter 3). Construction of the aqueduct commenced in 1886. It was 66 km long and comprised 40 km of open channel, 11 km of tunnels and 13 siphons, a total length of 13 km.[7] The original diversion weir on the river at Healesville would later be inundated by the Maroondah Reservoir. The aqueduct was completed in 1891 and named 'Maroondah' (Figure 1.9).[3] Prior to and during the construction of the works, all alienated land in the Watts River Catchment area was purchased by the Water Supply Board.[9] The entire town of Fernshaw was removed from its location adjacent to the Watts River on the Black Spur – even its cesspits were carefully excavated (Figure 1.10).[3] This was due to Davidson's insistence that the catchment area remain 'unalienated' (see Glossary for a definition). Indeed, Davidson said as much in a speech to the Royal Historical Society of Victoria in 1920 when he outlined the policy of the water supply branch of the Department of Public Works: 'that not a habitation or human resident should be permitted within any of the Melbourne water supply watersheds wherever situated. That policy, I think, is still persevered in'.[4] This policy is still strictly enforced in many parts of the network of Melbourne's water catchments, more than 120 years since it was first enacted.

Figure 1.9: The Maroondah Aqueduct on the Watts River opened in 1891 and provided a new supply of water to Melbourne (image: Melbourne Water).

Figure 1.10: The town of Fernshaw, on the Black Spur north of Healesville, was entirely removed from the Watts River Catchment area (image: MMBW[8]).

Continued growth of Melbourne

While the Maroondah diversion and aqueduct were being constructed, Davidson was identifying other water sources suitable to meet the demands of an expanding Melbourne. Following surveys of the southern slopes of the Great Dividing Range, he decided that the Upper Yarra River as well as two of its tributaries (the O'Shannassy River and Armstrongs Creek; Figure 1.11) would eventually be needed for water supply.[9] Davidson recommended to the government that the Upper Yarra Catchment should be reserved permanently for the purposes of water supply. This was notified in the *Victorian Government Gazette* in December 1888, and the O'Shannassy River and Armstrongs Creek catchments were notified in November 1889.[9] This was no small feat – the Upper Yarra Catchment alone was over 46 000 ha[12] and the idea of closing it to all activities was a rare proposition indeed. In Davidson's view, closure of the catchments was necessary to reduce pollution and contamination of water, and 'it is a question of the government resuming all land for water supply, or constructing such works as will prevent any pollution of the water in the reservoir from such settlement'.[14]

Figure 1.11: Armstrongs Creek (image: D. Blair).

Links between water quality, human activity and disease

The purpose of building the Yan Yean Reservoir was to secure a supply of clean water for Melbourne because the Yarra River had become increasingly polluted and its fetid water was a risk to human health. Earlier, Jackson's decision to position the diversion on the Plenty River downstream of the township of Whittlesea had meant that all the water going into the Yan Yean Reservoir was already contaminated by agricultural, industrial and human waste. As late as 30 years after the construction of the Yan Yean Reservoir, two farms still operated within the catchment and within a kilometre of the reservoir's shores.[14] Outbreaks of illness and disease, including typhoid, did not stop with the construction of Yan Yean Reservoir and this caused consternation among the citizens of Melbourne, who had thought that their health problems would be solved.

Box 1.7: Miasmas and germs – transmission of disease

Knowledge of how diseases such as typhoid are transmitted is fairly recent. From the beginning of medicine in ancient Greece to the middle of the 19th century, it was thought that many contagious diseases were acquired through contact with miasmas (polluted air). Malaria, for example, was so named (Italian: 'mal' = bad, 'aria' = air) because of the belief that it was contracted via miasmas or vapours from swamps, not from the bites of mosquitoes that lived in wet swampy areas. The idea that diseases may be transmitted directly between people, or via water, was simply unheard of.

Outbreaks of typhoid and its more virulent relative, cholera, were common in urbanised areas in the 19th century. People believed that noxious odours and miasmas from sewers, cesspits and other filthy areas were causing disease.[25]

Miasma theory was disproved in the late 1800s by the efforts of doctors and researchers who formed the basis of what we know now as the germ theory of disease. Through discoveries of the causative agents of contagious diseases (e.g. Robert Koch discovered that cholera was caused by a bacteria, *Vibrio cholerae*), how diseases were transmitted (e.g. John Snow showed that cholera was linked to drinking water contaminated by leakage from cesspits) and development of preventative measures to stop diseases (e.g. Louis Pasteur demonstrated the role of sanitation in preventing the spread of bacterial puerperal fever), it became clear that diseases were not generated spontaneously from bad smells. By developing separate systems for dealing with drinking water, and sewage and waste, the risk of contracting water-borne diseases like typhoid decreased. In Melbourne, typhoid levels declined from 13% per 1000 inhabitants in 1889 (560 cases in a population of 445 200) to 1% per 1000 inhabitants in 1902 (72 cases in a population of 501 585).[26] This was the result of the two-pronged approach of closing water catchments to human activities and developing a comprehensive sewerage system for the city.

Aside from the obvious problems with water contamination, the other main source of disease was the lack of an effective sewerage system in the city. At that time, the 'miasma' theory of disease was well established and people were warned to stay away from dirty, smelly areas for fear of contracting diseases from miasmas or 'bad air' (Box 1.7). During the 1870s and 1880s the field of microbiology developed as a science. A new understanding of the transmission of typhoid fever emerged when researchers in Europe identified a bacteria (*Salmonella enterica* var. Typhi) as the agent that caused the disease, and attributed its transmission to water contaminated with human waste, similar to that for cholera.[15,16] At the same time, the public health situation in Melbourne had become so bad that by the late 1880s, the government ordered a Royal Commission 'To inquire into and report on the sanitary condition of Melbourne'.[14]

The Royal Commission into the Sanitary Condition of Melbourne

The 1889 Royal Commission sought to examine the ways in which the water supply for Melbourne was affected by 'noxious trades' (e.g. abattoirs, gut factories, tanneries, night-soil depots, etc.) and to decide whether these trades needed to be removed from sites near watercourses.[14] Davidson, in his capacity as Superintending Engineer of Water Supply, provided evidence to the Commission that settlements close to watercourses in the Yan Yean system had the potential to pollute the water in the reservoir. The Water Supply Board had already bought out all the land-holders in the Watts River Catchment and the area had been reserved, temporarily at the time, for water supply.[14] Because of this, Davidson recommended to the commissioners that the entire catchments of Wallaby and Silver creeks and the East Plenty River be 'permanently reserved for water supply purposes', which included removing the small amount of sawmilling and timber-splitting from those areas.[14]

Some of the recommendations of the Royal Commission included:

> *that all alienated land within the direct catchment area of the Yan Yean Reservoir be resumed as soon as is practicable* (Recommendation 1, p. 19), and
>
> *that no milling or splitting licences be granted to cut timber on any lands reserved for the purposes of water supply, but not forming any actual catchment area* (Recommendation 19, p. 20).[17]

Perhaps most importantly, the commissioners did not limit reservation of land for water supply specifically to Yan Yean, also recommending:

> *that the conservation of all actual catchment areas in their integrity for purposes of water supply, and for no other purposes whatsoever, be guarded by the most stringent precautions* (Recommendation 18, p. 20).[17]

One commissioner, James Campbell, did not support the closure of catchments to the extent detailed in the recommendations. Making the point that 'danger from disease is not in the forest', which would not necessarily be true if the catchments were open, he held that the danger was 'in the reservoirs, the pipes, the streets and the lanes',[17] p. 20). This demonstrates that the link between catchment integrity and water quality was still not well understood.

A parallel aspect of the Royal Commission was the investigation of the drainage and sewerage of Melbourne, and of the quality of the supply of water. The commissioners recommended that a 'deep drainage system' for Melbourne would be required to combat the prevalence of diseases such as typhoid, in addition to the removal of sources of potential contamination from the water supply.[17] Following the recommendations of the Royal Commission on the water supply and sewerage systems, the Melbourne and Metropolitan Board of Works (MMBW) was formed in 1891.[5] This was the first time that water supply, sewerage, drainage and related infrastructure were under the control of a single organisation.

Thwaites was heavily involved with the design and development of the sewerage system. He presented a detailed proposal to the Royal Commission for the drainage and sewerage of Melbourne, based on his study of the topography and rainfall of the city. However, the tradition of using British experts still prevailed. Comprehensive plans were presented by Thwaites (who had been educated locally at the University of Melbourne) and others to the Royal Commission, but British engineer James Mansergh was nevertheless commissioned to come to Melbourne to assess the water supply and sewerage situation.[13] Mansergh examined the city and Thwaites' proposal, and eventually presented a proposal that was similar to the one submitted by Thwaites. However, Thwaites revised his original proposal and convinced the MMBW to adopt this scheme.[13] This scheme (called the M Scheme) would eventually become the sewerage system for Melbourne. It included the Spotswood Pumping Station and Werribee Farm for sewage treatment. It led to improved sanitation and reduced transmission of infectious diseases in water.

Conclusion

While Melbourne's water supply had been (relatively) reliable since 1857, the combination of open catchments and lack of sewerage infrastructure meant that much of the early history of the city was characterised by the effects of typhoid and water pollution. Following the new ideas of Blackburn and Jackson for supplying water to the rapidly growing population of the city, Davidson and Thwaites were responsible for bringing clean water from forested catchment areas and for removing waste to reduce contamination and the spread of disease. The high quality of the drinking water in Melbourne today is largely the result of the pioneering vision and expertise of these four men.

Chapter 2: 1901 to 1939

Introduction

Between 1901 and 1939, Melbourne's population grew from ~494 000 to almost 1 million.[27] New suburbs were established in the eastern and south-eastern areas, expanding the boundaries of the metropolitan area. This period was punctuated by several droughts such as those in 1907, 1914–15, 1919, 1922–23 and 1937–38, when water supplies were limited. Plans were developed to augment the supply of water to the city. The combination of droughts and population growth resulted in a flurry of construction, with three storage reservoirs built in 15 years. The bushfires of 1939 damaged much of the existing water catchment areas, which affected water yield in the 1940s, 1950s and beyond. These major events feature in this chapter.

'Enemies of water supply': drought and rapid population growth

A drought in 1907 and 1908 placed considerable pressure on Melbourne's water supply. The city was growing, with a population of over 530 000, and development was increasing in the eastern suburbs.[7] People were building houses on large blocks with substantial gardens, which meant that average water use was 270 L per person per day.[1,5] By comparison, water usage for the 2011–12 financial year was ~142 L per person per day.[28] The increase in population during the early 1900s, together with the need to supply water to houses at higher elevations than the inner suburbs at sea level, placed additional pressure on the water supply system. The Maroondah Aqueduct, completed in 1891, supplied 127 ML per day to the Preston Reservoir.[1] In 1908, this aqueduct was supplying almost 70% of the city's water, far more than its usual contribution. This was because the drought had caused Yan Yean Reservoir to fall to a depth of 3.6 m, its lowest ever.[1,5] It was clear that other sources of supply were needed. Where was more water going to come from?

The Acheron option

The Acheron River, a tributary of the Goulburn River north of Marysville, had been investigated as a potential water supply in the late 1890s (Figure 2.1). Thwaites had calculated that the river could be diverted fairly cheaply, for around £50 000 in 1901 values.[5] The MMBW gave testimony to an interstate Royal Commission on the condition of the Murray River,[29] arguing that it should be entitled to divert water from the Acheron River at all times except January to April inclusive (i.e. the summer peak periods when streamflow is low). The MMBW requested a volume of no more than 90 ML per day.[26] However, the MMBW's proposal was not acted upon by the state government. Requests for 5180 ha of the Acheron River watershed to be reserved for water supply, made as part of the MMBW's testimony to the Victorian Royal Commission on State Forests and Timber Reserves (which ran between 1896 and 1901) were rejected by the state government in 1900. This was because water flowing to the northern side of the Great Dividing Range, as does the Acheron River, had traditionally been viewed as water for the Goulburn Valley farmers.[9] Opposition to Melbourne 'taking' the Goulburn Valley's water would reappear every time such a plan was proposed. During the 1907–08 drought, however, the state government approved the diversion of water on a temporary basis from the Acheron

Figure 2.1: Acheron River (image: D. Blair).

River into the Maroondah Catchment to augment supply of water to Melbourne,[5] provided the MMBW paid for the water. The MMBW was not happy with this proposition. The diversion from the Acheron River did not proceed and the existing supply only just made it through the drought.

New water supplies: the O'Shannassy River and augmentation of the Maroondah system

Following surveys of the Upper Yarra region in 1901, the MMBW recommended the construction of a diversion weir and aqueduct on the O'Shannassy River. This could supply 113 ML daily to Melbourne via a new service reservoir to be built at Mitcham.[26] The MMBW suggested that the Upper Yarra River also could be diverted into the O'Shannassy Channel, which would double the amount of water available to Melbourne each day.[26]

In late 1907, the MMBW surveyed the O'Shannassy River and decided that a diversion of the river was a better proposition than a new storage reservoir on the Watts River downstream of the existing diversion weir for the Maroondah Aqueduct.[5] At the same time, the Forests Commission of Victoria lobbied the government for a Bill to be introduced to give control of the O'Shannassy watershed to the Forests Commission.[5] Foresters argued that timber industries and water supply could co-exist, and the government at the time was opposed to MMBW's policy of closing catchment areas.[5] The government refused to have the land vested in the MMBW, despite the land being reserved for water supply since Davidson's recommendation in 1888 (with the O'Shannassy Catchment reserved in 1889).[9] However, a narrow strip of land on either side of the O'Shannassy River was offered for vestment to the MMBW.[5] The MMBW had no choice but to accept this offer, but reapplied in 1908 to have the whole O'Shannassy and Upper Yarra rivers watersheds vested – formalising Davidson's request for the areas to be designated for the sole activity of supplying water. Davidson's ongoing influence in keeping catchments unalienated was evident in 1909 when, as part of a Board of Inquiry into the potential developments of the O'Shannassy and Maroondah systems, he stated:

> *the purity of the water supply to be of much greater importance than the utilization of the belts of useful timber, and we therefore unhesitatingly recommend that the area* [the O'Shannassy watershed] *be permanently reserved for water supply purposes, and for water supply purposes only* (Board of Inquiry to Water Supply,[30] in Seeger[9]).

The government vested the O'Shannassy River Catchment to the MMBW in 1910.[1]

Debates over forests aside, questions about sources of additional supplies of water still remained. Information gathered by the MMBW had shown that the O'Shannassy and Upper Yarra regions could provide many opportunities for water harvesting and storage. Davidson had predicted, during the construction of the Maroondah system, that a permanent water storage would be needed on the Watts River.

The Board of Inquiry investigated the options of diverting the O'Shannassy River or augmenting the Maroondah system.[9] It handed down its findings in 1909. The main conclusion was that the O'Shannassy River should be utilised ahead of any augmentation of the Maroondah system.[30] The justification was simple – not only could the O'Shannassy River supply more water to a larger population at a lesser cost, it also would be a separate system from both Maroondah and Yan Yean and thus provide greater security to the water supply for the city.[1,9] Both the Maroondah and the Yan Yean systems supplied the Preston Reservoir north of Melbourne for distribution to the city. By contrast, the O'Shannassy scheme could run to the east of the city.[26] An additional benefit of the O'Shannassy scheme was its elevation. The proposed site of the diversion weir on the O'Shannassy River was higher than the site of the proposed Maroondah storage, enabling a more consistent gravity-fed supply of water to the developing (and higher elevation) eastern and south-eastern suburbs of Melbourne. Augmentation of the

Box 2.1: Land for water supply – reserved v. vested

The government of Victoria had the power to allocate Crown land as it saw fit. Allocating land for the supply of water to Melbourne was controversial, and arguments erupted frequently as parts of the Yan Yean Catchment were closed to activities such as logging and farming (see Chapter 1). While the O'Shannassy River and Upper Yarra River catchments had been reserved for water supply in 1888 on Davidson's recommendations, they were not vested in the MMBW at its inception in 1891.[5,9] The difference between reserved and vested land is that all activities were banned on reserved land. For water supply infrastructure to be developed, the reserved land needed to be vested into the MMBW's possession. The MMBW would then be in total control of the land usage, which is why applications for land to be vested in the MMBW were resisted by foresters and others with interests in potential land uses. In the catchments of the O'Shannassy and Upper Yarra rivers, no plans could be implemented until the lands were vested in the MMBW. This made the development of new water supply options difficult.

Maroondah system would require pumps to supply those suburbs – an expensive proposition.[7]

New water supplies: the O'Shannassy River

The construction of a diversion weir on the O'Shannassy River and a 77 km system of aqueducts and siphons to convey 90 ML of water daily to the newly completed Olinda Service Reservoir (instead of the Mitcham Reservoir, as proposed by Thwaites) in the Dandenong Ranges took place between 1911 and 1914.[7] From Olinda Reservoir, the water travelled via pipeline to a service reservoir at Surrey Hills for distribution in the reticulated supply. The new aqueduct and service reservoir scheme was designed to provide back-up against potential breakdowns in the Yan Yean and Maroondah schemes. Furthermore, the positioning of new aqueducts, pipelines and service reservoirs in the east of Melbourne facilitated the supply to the rapidly expanding southern and eastern suburbs (Figures 2.2, 2.3).

Figure 2.2: a) The O'Shannassy Aqueduct carried water from the diversion on the O'Shannassy River to the Olinda Service Reservoir (image: Public Record Office Victoria). b) The O'Shannassy Aqueduct, present day (image: D. Blair).

Figure 2.3: Wooden stave pipes in the O'Shannassy Aqueduct (image: Australian Wood Pipe Company[31]).

Box 2.2: Edgar Ritchie – visionary of water supply

Edgar Ritchie (1871–1956) was born in Melbourne, and began his career in water supply and engineering as a trainee draughtsman in the Victorian Department of Public Works in 1888.[45] He worked in the water supply department of the MMBW after its formation in 1891, and was promoted to Head Engineer of Water Supply in 1908. Ritchie oversaw the construction of the Maroondah, O'Shannassy and Silvan reservoirs and their associated infrastructure. Ritchie also conducted comprehensive surveys of land for future potential water supply.[45] He was instrumental in the expansion of the water catchment areas for Melbourne, having had the O'Shannassy and Upper Yarra River catchments vested in the MMBW during his time at the helm of the Water Supply Department.[5] Ritchie also was responsible for the proposal to utilise water from the Thomson River (see Chapter 4). Although he was unsuccessful in having the Thomson Catchment vested, it would later feature as the centrepiece of Melbourne's water supply. Ritchie's work built on that of Davidson, and the actions of these two men ensured that Melbourne had high-quality water. Ritchie retired from the MMBW in 1936.

Edgar Ritchie (image: State Library of Victoria).

Drought in 1914 made it obvious that additional water supplies were required, even with the new O'Shannassy diversion in operation. Edgar Ritchie (Box 2.2), then Head Engineer of Water Supply at MMBW, examined the overall state of Melbourne's water supply in 1915. He recommended the construction of a dam on the Watts River downstream of the Maroondah diversion, and the augmentation of the O'Shannassy system with construction of a dam on the river, upstream of the existing diversion weir. The MMBW gave authority to proceed with preparations of potential designs of the dam at Maroondah in 1917, and conducted preliminary land-clearing at the site of the reservoir.[32]

1920s: further expansion of water infrastructure

Maroondah Reservoir

Work on the Maroondah Reservoir commenced in 1920 and was completed in 1927. Water from the new reservoir would flow down the existing aqueduct system to the Preston reservoirs. In conjunction with the dam project, the capacity of the existing aqueduct was increased and the siphons were duplicated to accommodate the increase of flow once the new reservoir was operational.

Box 2.3: Gatekeepers of water – the water supply caretakers

Much of the water destined for Melbourne flowed along open aqueducts from forested catchments to metropolitan storage facilities. These aqueducts were a key part of the Maroondah, O'Shannassy and Silvan systems. The aqueducts followed the contours of the land and, where they crossed gullies, culverts were built under the aqueduct to allow water to run under (rather than into) the aqueduct. These networks, together with the older aqueducts servicing Yan Yean Reservoir, required the constant attention of caretakers positioned along their length.

Caretakers and, to a lesser extent, their assistants were responsible for sections of the aqueducts, usually 8–16 km in length. In most instances, a caretaker lived on site with his family in a house provided by the MMBW.

Aqueducts were a vulnerable component of the water supply system because they were subject to potential slips, blockages and damage from falling trees. Caretakers monitored the entire length of the aqueducts, ensuring that water flowed freely and there was no leakage. The challenges to supply went beyond land slips and leaks caused by heavy rainfall. For example, burrowing by a Common Wombat (*Vombatus ursinus*) destabilised the O'Shannassy Aqueduct and caused a major washout in 1946 (R. Kermode, pers. comm.).

During storms, the caretakers would be out in the rain even during the night, clearing fallen branches from gratings to ensure that the water supply was not compromised. Another key responsibility was to listen for, and detect, leaks. Tracks ran along the entire length of the aqueduct, on the downhill side. Caretakers walked the 'bottom track' once a week to detect any leakage. Leaks could be seen running from above or across the bottom track, and water could be heard trickling from the aqueduct. Leaks were costly from a water supply point of view and had the potential to cause total failure of the aqueduct. This occurred in 1923 at Block Cutting and in 1946 at Millgrove, on the Dee section of the O'Shannassy Aqueduct. The latter was repaired by building a steel flume supported by timber poles, not unlike Davidson's timber flume at Yan Yean in 1878 (see Chapter 1). The caretakers or the local MMBW district workforce would repair leaks, ensuring that the water moved freely through the system and reached its destination.

Caretakers were also stationed at every storage reservoir and major metropolitan service reservoir where they monitored water levels, adjusted flows and assisted with the movement of water around the supply system. Instructions to caretakers about water supply operations were originally sent by telegraph in written form. This changed with the introduction of the magneto telephone line. Communications were further improved with the introduction of the radio telephone at management offices (at Mitcham, Preston, Warburton, Yan Yean and Healesville) and in motor vehicles.

Between the 1940s and 1960s, caretakers received only one weekend off per month. As caretakers were required to be available 24 hours a day, some had alarms under their beds that would ring if there was an emergency. The union took up this issue with management but, when advised that watershed engineers also had bells under their beds, the union withdrew its complaint. Caretakers never received overtime pay – irregular and long hours were part of the job. Labourers from the local workforce covered periods of leave but were not paid extra for their additional temporary responsibilities. They were usually told that they should be grateful that they were given the opportunity to work as a caretaker, as it involved the potential for future promotion.

In the 1940s, assistant caretakers were appointed to each station, giving the caretaker an additional day off each month. Conditions for caretakers were improved significantly in 1970 with assistant caretakers assuming increased responsibility by working 10 day fortnights and sharing out-of-hours work. Their salaries were adjusted to include allowances for overtime and out-of-hours availability.

Transport for caretakers changed over the years. Initially they moved around on foot, then used bikes for most of the time between 1930s and 1960s (they received a bike allowance as part of their salary). In the 1970s and 1980s they were supplied with cars and the length of aqueduct for which they were responsible increased.

Figure 2.4: Construction of the Maroondah dam wall with aerial ropeway in the background (image: Public Record Office Victoria).

Figure 2.5: Boulders being set in concrete for the Maroondah dam wall (image: Public Record Office Victoria).

The original Maroondah Aqueduct system included several diversions from other creeks. Graceburn, Donnellys and Sawpit creeks were diverted into the Maroondah Aqueduct downstream of the diversion weir between

Figure 2.6: Maroondah Reservoir and outlet tower (image: D. Blair).

1886 and 1893. Coranderrk Creek was diverted into the Maroondah Aqueduct in 1909. This last diversion was connected via a pipeline into the Graceburn Aqueduct, which in turn fed into the Maroondah Aqueduct.[33] In 1929, a further weir and aqueduct diverted water from Coranderrk Creek into the O'Shannassy system.[34]

The location of the Maroondah Reservoir made transporting materials and equipment difficult. A 3.6 km aerial ropeway was installed between the railway station at Healesville and the dam site. The ropeway comprised 84 buckets powered by a steam engine,[35] and was used to haul the cement and sand required for the concrete dam wall. Large boulders were placed into the concrete wall to provide stability and reduce the amount of concrete used (Figures 2.4–2.6).[7]

O'Shannassy Reservoir

Construction of the O'Shannassy Reservoir began in 1923 after a drought in 1922–23 highlighted the urgent need to increase the water supply. Work proceeded under the supervision of resident engineer Alexander Kelso, who later became the Head Engineer of Water Supply following Ritchie's retirement in 1936.[5] The O'Shannassy Aqueduct was increased in size to provide for the larger volume of water that would flow along it from the new reservoir. The MMBW put out calls for tenders, which were not successful. Therefore, unlike the Maroondah Reservoir which was built by contract

Box 2.4: The Warburton Roar

Construction of O'Shannassy Reservoir was completed in 1928. It was unique as the spillway was five-stage siphonic, the stages cutting in at different levels as the level of the reservoir rose. The discharge rate of each siphon was 2.2 GL/d, giving a total potential spillway discharge of 11 GL/d.

The average annual inflow into the dam is 105 GL and the storage capacity of the reservoir was only 3.12 GL.

The spillway only operated twice. The first time was during the flood of 1934 but it is not known how many of the spillway siphons stages cut in at that time. It is legend that the noise was loud enough to be heard at Warburton more than 15 km away! The other occasion was when Ritchie asked to see it operating. When the first stage cut in he is reported to have asked, 'How do we stop it?', to which the response was, 'We can't, we have to wait till the reservoir level falls for the siphon to break.'

When remedial works were carried out on O'Shannassy Reservoir in the early 1990s to meet modern design standards the siphonic spillway was removed and replaced with an overflow spillway. The dam is now allowed to spill.

Figure 2.7: a) Construction of the O'Shannassy dam wall. b) The completed reservoir (images: Public Record Office Victoria).

Figure 2.8: A reminder of how the Commissioners' Quarters functioned remains today – a flat stretch of land where the manager kept cows to provide fresh milk and cream for MMBW Commissioners and their guests (image: D. Blair).

under MMBW supervision, the MMBW itself built the dam at O'Shannassy.[5] This required the purchase of heavy earth-moving machinery and concrete-making equipment, and employment of large numbers of men as day-labourers.[5]

The O'Shannassy Reservoir is an earthen embankment with a reinforced concrete core, with additional puddle-clay on the upstream side (Figure 2.7).[7] At the time of construction, it was the highest earthen reservoir embankment in Australia.[7] Workers were housed in tents in one of the wettest parts of Victoria and conditions during construction were trying, especially in winter. This, and other problems such as pay disputes, caused disharmony among the workforce.[5] The O'Shannassy Reservoir was completed in 1928, a year after the Maroondah Reservoir.

Downstream of the O'Shannassy Reservoir, a large building was constructed to provide accommodation for the commissioners of the MMBW, who could be allocated a weekend or mid-week stay. The building was known as the Commissioners' Quarters, and its most high-profile guest was Queen Elizabeth II, who stayed there during a visit to Victoria in 1954 (Figures 2.8, 2.9). As it was downstream of the reservoir, the presence of livestock and

Figure 2.9: In 1954, Queen Elizabeth II and Prince Philip, the Duke of Edinburgh toured Australia. One stop on their itinerary was the Commissioners' Quarters at the O'Shannassy Reservoir (image: Melbourne Water).

buildings posed no threat to the integrity of the water quality at O'Shannassy Reservoir. There were similar buildings for commissioners at Wallaby Creek, which were destroyed in the 2009 Black Saturday bushfires. The O'Shannassy Commissioners' Quarters still exist and are now privately managed.

Silvan Reservoir

Ritchie identified two sites for off-river storages at Silvan near Mt Dandenong[7] in 1917. The proposal to construct reservoirs at this location was kept secret, as it would require purchase of alienated land. A Crown grant in 1919 allowed the MMBW to buy the land from its owners and have the area vested.[5]

Figure 2.10: A steam shovel in action during construction of the Silvan Reservoir (image: State Library of Victoria).

Construction of Silvan Reservoir took place between 1926 and 1932.[7] Learning from the problems associated with poor working conditions at O'Shannassy Reservoir, the MMBW built accommodation huts and service buildings for workers.[5] The widespread use of machinery also eased conditions for workers, with steam shovels, locomotives and steamrollers used to complete the work (Figure 2.10). Night shifts were made possible by generator-powered floodlights.[5,36]

The wall of Silvan Reservoir was constructed across Stoneyford Creek upstream of its confluence with Olinda Creek, a tributary of the Yarra River. The wall is a rolled earthfill embankment with a 1.2 m thick reinforced concrete core,[36] similar to the O'Shannassy dam wall. An innovation on the Silvan dam wall was the inclusion of a drainage system on the downstream side to protect against failure. A series of vertical wells from the open edge of the wall to the natural floor of the valley was used to divert water that might otherwise enter the wall via shrinkage cracks in the earthfill.[36] Another innovation was the duplication of the outlet structures, allowing inspection and maintenance to take place without having to shut off the supply of water.[36]

The Silvan Reservoir was designed to store water from the O'Shannassy Reservoir, Coranderrk Creek and the planned Upper Yarra schemes.[2] It remains a vital part of Melbourne's water storage system: although the O'Shannassy Catchment is highly productive, with an annual average yield of 105 GL, the O'Shannassy Reservoir has a storage capacity of only 3.1 GL. Therefore, to maximise harvest, water must be transferred quickly to Silvan Reservoir for storage and subsequent distribution to the city.

Construction of the new reservoirs at O'Shannassy, Maroondah and Silvan required augmentation of the existing city mains and service reservoir system. The new reservoirs increased the ability of water engineers to provide adequate supplies to the expanding south-eastern suburbs, via manipulation of the aqueduct and supply system. Construction of many additional service reservoirs, such as at Notting Hill, Waverley, Research, Surrey Hills and Mitcham,[5,32] and the establishment of pipelines between them allowed storage of water close to consumers, provided for daily demand and ensured consistent supply and pressure to households.

It was around this time that greater interest was taken in water quality. Regular comparative samples of water were collected from houses at the end of the reticulated supply and from the headworks of the main reservoirs to ensure that the water retained its quality as it moved along the system.[32]

Ritchie's ideas for Melbourne's water and securing of supply options

Edgar Ritchie had visionary ideas for the water supply for Melbourne. In 1915, along with recommending the construction of the Maroondah and O'Shannassy reservoirs, he had visited the source of the Upper Yarra River to determine the possibility of diverting the headwaters of the Thomson River into the Upper Yarra Catchment. Following this, he proposed that an aqueduct from the Upper Yarra River and Armstrongs Creek could be built to divert water into the O'Shannassy Aqueduct to supply water for 100 000 extra Melburnians.[5]

In 1922, Ritchie produced a report on the state of Melbourne's water supply.[37] He made several proposals that would form the basis of the majority of new capital works for supply of water to Melbourne over the next 60 years, including diverting water from Armstrongs Creek.[7] This built on the idea put forward by Davidson in 1888, that the creek could be diverted easily into the Silvan system, which (at the time) had not been built although

Box 2.5: Fishing in the water supply reservoirs

Edgar Ritchie was a very keen fly fisherman, and would occasionally be driven up to fish for trout at the Toorourrong Reservoir. He reportedly would not let his driver fish, but would instead give him the first, third and fifth fish that he caught (R. Kermode, pers. comm.). Today, fishing is strictly forbidden in water supply reservoirs, but special permits were issued to allow restricted fishing in some of the MMBW's reservoirs until 1946. Commissioners and MMBW employees could apply for permits to fish for trout in Toorourrong Reservoir or the newer O'Shannassy and Silvan reservoirs. The Silvan Reservoir was the site of a trout hatchery. Yan Yean was not included as there were no trout in that reservoir. Successful applicants would have to present themselves at the reservoir gates on their allotted day and the caretaker would let them in for their day of fishing. Of course, various opportunistic and surreptitious fishing events were also conducted in reservoirs by workers and contractors.

plans for it had been developed. While Davidson had succeeded in having these areas reserved for water supply purposes, the MMBW did not have the land in the Upper Yarra River Catchment vested.

Ritchie had established stream gauges to measure water yields in Armstrongs Creek and he proposed several small storage dams on the Yarra River to hold the water.[5] On his recommendation, in 1923 the MMBW applied to the government to vest, for water supply purposes, the whole of the Upper Yarra River Catchment upstream of the O'Shannassy River, the Baw Baw Plateau and the catchments of the Thomson and Aberfeldy rivers.[7] Ritchie calculated that schemes on the Armstrongs Creek, Upper Yarra River and Baw Baw systems would provide enough water for a city of 3 million people.[7] He surmised that it would take 50–60 years for Melbourne's population to reach that level. Following that, he projected for the additional population beyond 3 million people, the Thomson and Aberfeldy rivers would need to be tapped.[37]

Like his predecessor (Davidson), Ritchie was staunchly opposed to any human activities in water catchment areas. Having described the practices of cattle graziers and the sawmilling industry as 'a crass folly of forest destruction',[38] he often came up against the government and the Forests Commission of Victoria. In his opinion, large trees should be protected instead of logged, as they afforded great gains in water runoff in catchment areas.[39] Ritchie held views well ahead of his time, particularly on the relationships between large trees, water yield and quality, and increased streamflow and reduced erosion. Subsequent research has demonstrated that runoff from areas of regenerating Mountain Ash (*Eucalyptus regnans*) forest is markedly lower than runoff from mature old-growth forest[40,41] (see Chapters 4 and 5).

Ritchie was particularly scathing of the way alpine lands were treated by cattle graziers, who regularly burned stands of Snow Gum (*Eucalyptus pauciflora*) to promote grass growth to feed livestock.[38] This was relevant to the MMBW's application to have part of the Baw Baw Plateau vested for water supply purposes. However, in 1928 the MMBW failed in its attempts to have vested the Baw Baw Plateau, and the Thomson and Aberfeldy catchments. Instead, the government allowed the MMBW exclusive use of half the Upper Yarra Catchment area it had requested (18 210 ha instead of 36 421 ha), in an agreement with the Forests Commission of Victoria.[34]

Commencement of the Upper Yarra scheme

In 1929, following reservation of the land around the Upper Yarra River and catchment area, construction began on an aqueduct and siphons from a diversion weir on the Upper Yarra River to the O'Shannassy system.[34] It was also around this time that the decision was made to build one large dam on the Upper Yarra River, instead of the series of small reservoirs as initially

Figure 2.11: Regrowth from the 1926 bushfires at O'Shannassy and Maroondah, showing the densely packed young trees (image: E. Beaton).

proposed by Ritchie.[34] The aqueduct took nine years to complete as its construction was slowed by the Great Depression (1929–32). During this time, survey parties examined potential sites for the new Upper Yarra Reservoir, laying the groundwork for development of the dam in the 1950s. Augmentation of the water supply scheme would be necessary as Melbourne's population had grown to over 994 000 by 1939. Water usage was continuing to increase, with ~295 L used per person per day in 1939.[42]

Bushfires: 1939 Black Friday fires

Bushfires in January 1939 (later called the Black Friday bushfires) burned ~65% of the Maroondah Catchment and substantial parts of the Upper Yarra Catchment. Only limited fire damage occurred in the O'Shannassy Catchment.[5] Entire towns in the Yarra Ranges and Central Highlands were destroyed (e.g. Narbethong and Woods Point) or extensively damaged (e.g. Healesville and Marysville).[43] Earlier fires in 1926 in the Maroondah

Box 2.6: Surviving the 1939 Black Friday bushfires

Neville Smith was the Director of Engineering (1982–85) and Deputy General Manager (1985–86) of the MMBW. In the 1930s, his family lived in a tin-roofed hut at Icy Creek, located between Noojee and Tanjil Bren, and survived the 1939 bushfires. As the fires reached their home, Smith's father used long chains (normally used at the Warburton sawmill) to tie down the roof of the cottage to survive the coming firestorm. Smith was five years old at the time and he remembers seeing the fire coming over the hill towards the valley where he lived. He was not scared because he had an older brother whom he believed would protect him from anything, including fire. Almost all the houses in the village were destroyed, except the Smiths' hut. That night, the house was filled with people sleeping on the floors in all the rooms, as they had nowhere else to go.

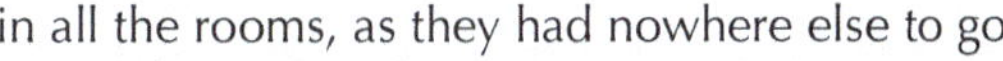

Neville Smith's grandfather lived nearby with his family. When the fire came they ran from their house into a ploughed and cleared potato field, throwing wet woollen blankets over themselves for protection. However, Smith's grandfather was not with his family. After the fire they found him walking over from another paddock. He'd run in the opposite direction from the family to save his horse. He and the horse had huddled under a wet blanket and both had survived. Smith remembers the horse well, which reportedly lived into its 30s.

MMBW Chief Engineer Neville Smith (image: Neville Smith).

Catchment had resulted in tree regeneration (Figure 2.11),[44] but the intense fire in 1939 killed those immature trees and there was very little chance of natural seed germination and forest regeneration. Consequently, parts of the Black Spur area in the Maroondah Catchment were replanted by hand with rows of Mountain Ash.

Conclusion

The period between 1901 and 1939 was characterised by an increase in demand for water, additional infrastructure to meet that demand, and increasing requirements for high water quality. The MMBW worked hard during that time to manage water infrastructure and meet the various demands.

From Jackson's original projection for Yan Yean to supply 200 000 people each with 136 L of water per day (see Chapter 1), the usage of water had increased to over 290 L per person every day, in a city of almost a million people. Ritchie's foresight in understanding forest ecology and how it related to water supply, his designing of practical solutions for water supply and his accurate projections of supply and demand assisted the MMBW's decision-making for most of the 20th century. Ritchie's insistence on a closed water catchment policy ensured continuation of the improved water quality for Melbourne that had begun with the efforts of Davidson and Thwaites.

Chapter 3: 1940 to 1960

Introduction

Melbourne's population grew beyond 1 million during the 1940s.[27] The effects of the 1937–39 drought were still felt keenly in the early 1940s, despite the significant augmentation of the water supply system during the 1920s. Drought struck again in 1943–44,[46] reinforcing the idea that more options were needed to maintain a secure water supply and protect the growing population from water scarcity. In this chapter, we explore the effects of the 1939 bushfires on the catchments and the development of the Upper Yarra Reservoir – a new water supply for Melbourne.

Aftermath of the 1939 bushfires

By the time of the 1939 bushfires, there was evidence of links between the quality and yield of water and the integrity of vegetation in the catchments. Management of fire was a new facet of the ongoing activities of the MMBW. Following the bushfires, Chairman of the MMBW, John Jessop, stated in an address in 1942:

> *The adequacy of the supply depends in no small measure on the retention of the pristine characteristics of the watersheds, the greatest menace to which is fire. The bitter lessons of the tragic bushfires of 1939 have not passed unheeded and it is now accepted that, if the areas are to continue to yield abundant supplies of water, greater protection from such fires will have to be afforded. The Board, however, is mindful of the fact that only a relatively small portion of the forests in this State are under its control, and realises that if any fire control policy is to be effective co-operation with the Forests Commission is essential* (p. 26).[20]

Large areas of forest were burned during the 1939 Black Friday fires, both within and outside the water catchments. A swift but extensive Royal

Figure 3.1: MMBW worker manning a fire tower (image: Public Record Office Victoria).

Commission[47] into the cause of the fires and potential prevention of future fires was conducted. The MMBW was questioned about its fire management practices for the forests under its control. The long-standing feud between the MMBW and the Forests Commission of Victoria over land use was noted in its report, and the MMBW policy of not burning in catchments was criticised by the commissioners. They speculated, based on evidence given by the MMBW and by land-owners and other parties, that the condition of the land (i.e. catchment areas with no active fire hazard reduction strategies) within the MMBW's authority may have contributed to the loss of buildings and lives during the bushfires. The commissioners reasoned that the spread of fire may have been contained if regular preventative burning in catchment areas had been conducted.[47] The MMBW policy of not conducting regular burning in water supply catchments was based on the understanding that prescribed burning was inappropriate for use in Mountain Ash and Alpine Ash forests because fire is often a stand-replacing form of natural disturbance in those vegetation types, and fire affects the quality and quantity of streamflow in catchments (see Chapter 5).

Figure 3.2: The Mt St Leonard fire tower in 1940 (image: Healesville Historic Society).

Fire management and fire towers

A key outcome from the Royal Commission into the 1939 fires was the directive that the Forests Commission of Victoria and the MMBW be required to erect lookout towers (called fire towers by the MMBW) to detect fires before they burned forested areas. These towers were to be connected by telephone and would be operated over summer, the peak season for fires.[47] Construction of MMBW fire towers began in the early 1940s, with the first built at Mt Poley in the O'Shannassy Catchment. The towers were strategically placed on top of mountains to enable watchmen to see as far as possible. The MMBW also built towers at Strath Creek (in the Wallaby Creek Catchment), Mt St Leonard (near Healesville, for the Maroondah Catchment) and McVeighs (Upper Yarra Catchment). On a clear day, the Mt St Leonard tower has a view as far as Geelong – more than 140 km away (Figures 3.1, 3.2).

Another development following the 1939 bushfires was the employment of foresters in the MMBW workforce. The MMBW foresters were responsible for protecting the integrity of the forested catchment areas via fire

Box 3.1: Fire towers

The first fire tower was built and manned in the early 1940s in the O'Shannassy Catchment by Gordon Sumner, a labourer with the MMBW at Warburton. Sumner built the tower in a large Mountain Ash tree on which he constructed a platform, using ropes and pulleys to drag materials and equipment up the tree. Perhaps more remarkable was the fact that Sumner completed the construction and tree-climbing with a prosthetic leg. His leg had been amputated below the knee as a young boy. Sumner manned his tower for 12 out of 14 days, riding a bicycle from the O'Shannassy caretakers' quarters to the tower.

The MMBW constructed several other fire towers and employed 'tower men' every summer. Following notification of a fire, crews from the Country Fire Authority and the Forests Commission of Victoria responded to fires on farmland or other areas. The MMBW had fire-fighting equipment and used its network of roads to access fires burning within the catchments.

Towers were manned all summer. They were occupied each day from early in the morning until late in the evening, or even overnight if it was a stormy or hot night. The MMBW towers were renowned for being manned longer than those of other organisations. Generally, there wasn't much to do while up a tower looking for fires and, apparently, some of the men passed the time by sunbaking in the nude (D. Lindenmayer, pers. obs.).

prevention and maintaining security of the closed catchments. The forestry team constructed roads throughout the catchments to enable access for fire fighters. The road network was comprehensive, with no point within a catchment more than 1.6 km from an access track. Markers along the numbered tracks displayed the distance from a fixed point to help guide people and pinpoint the location of fires.

Teams of workers led by foresters were responsible for the maintenance and protection of the catchments. Forest gangs worked year round, removing bracken, clearing and building roads and conducting general maintenance of the area. Foresters were also responsible for supervising the MMBW fire-fighting team, as the catchment lands did not fall under the jurisdiction of the Country Fire Authority or the Forests Commission (the latter eventually became part of the state Department of Natural Resources and Environment). Because the catchment areas were vested in the MMBW they were considered metropolitan land, despite being far from the limits of the metropolitan border (F. Lawless, pers. comm.).

The MMBW forestry team had considerable expertise and equipment at its disposal to fight fires. Fire fighters came from within the MMBW – approximately 100 workers enlisted for training and were able to assist when the alarm was raised.

Box 3.2: Lightning strikes in forests

Most fires in the catchments are started by lightning. A key part of fire suppression is putting out the fire as quickly as possible, so it was important to find the trees that had been hit and stop them from burning (F. Lawless, pers. comm.). Trees hit by lightning burn slowly from the inside, giving fire fighters time to locate them before the fire can spread. The challenge was to find a smoking tree in a heavily forested area. The MMBW network of fire towers enabled spotters to use compass bearings to pinpoint the position of a lightning-struck tree. Once trees struck by lightning were mapped, the distance to the nearest access road could be estimated. Teams of fire fighters would be directed to the relevant access road, with instructions on how far along the road to travel to the point closest to the tree on fire. They would be given a compass bearing along which to walk from the road into the forest at the designated point. This approach proved to be highly effective.

Fire-fighting requires special equipment; for example, regular industrial bulldozers are unsuitable for constructing firebreaks. To create a firebreak, a forestry bulldozer with an angled blade and a tree-pusher is required instead of a flat-bladed construction bulldozer. The MMBW invested hundreds of thousands of dollars in developing specialised fire-fighting infrastructure to protect the catchments (F. Lawless, pers. comm.).

Integrity of the water supply: protection from disease

As discussed in Chapter 1, Yan Yean Reservoir was constructed as an alternative water source to the increasingly polluted waters of the Yarra River. However, typhoid fever still killed hundreds of people in Melbourne even after the development of the Yan Yean scheme[26] (see Box 1.7). While poor water quality was mainly due to an ineffective (or absent) sewerage system, the presence of humans, livestock and industry inside the catchment also significantly reduced water quality.

Closure of the Yan Yean and Watts River (Maroondah) catchments resulted in a decline in deaths from typhoid. The reduction in mortality rates was also due to the introduction of a sewerage scheme which completed the cycle of separating drinking water from potential sources of pollution. The decline in diseases set a precedent for future catchment closures for the O'Shannassy and Upper Yarra regions. As outlined in Chapters 1 and 2, both Davidson and Ritchie fought to have catchment closure recognised as integral to the quality of proposed water supplies.

An outbreak of typhoid among timber workers near Warburton in the late 1920s was a strong reminder of the risks of contaminated water supplies and a vindication of the value of the closed catchment policy.[5] By the 1940s, it was discovered that people could carry and transmit typhoid bacteria while remaining asymptomatic.[48] The MMBW commenced blood testing of all employees working in the catchment areas; if results were inconclusive, stool samples were taken for additional testing.[20,49] MMBW forester Frank Lawless and MMBW engineer Frank Barnes remember instances of people revealed by the testing as asymptomatic carriers of the typhoid bacteria and being removed from working in the catchments. Book author Jim Viggers was blood tested when he commenced work with the Watershed Department of the MMBW in 1967. The testing continued into the 1980s, with book co-author David Lindenmayer tested before being allowed into the catchments to study Leadbeater's Possum (*Gymnobelideus leadbeateri*).

Expansion of the Upper Yarra scheme: the reservoir

Work on the first diversion of water from the Upper Yarra River commenced in 1929, with an aqueduct and pipeline system constructed to send water to the O'Shannassy Aqueduct[34] (see Chapter 2). Some of the land Ritchie had recommended for the Upper Yarra Catchment had been reserved for water supply since 1928, but the land was not vested in the MMBW and so the MMBW did not have total control of the catchment (Figure 3.3). Instead, it had an agreement with the Forests Commission of Victoria and the state

Box 3.3: The hole in the wall

Challenges to water supply also came from infrastructure. MMBW engineer Roy Kermode remembered how, in 1948, Yan Yean Reservoir overflowed for the first time in many years. The Engineer for Water Supply, F.M. 'Dickie' Lee, had wanted to see the overflow, so was driven out to observe it. While he and several others were out on the dam wall, Lee thought he could see a leak in the dam wall on the downstream side. He was assured that it was not the case, that Yan Yean's wall was impenetrable and that there was no way there could be a leak. However, later that day the caretaker went back to examine the area and there was indeed a leak in the wall. Using picks and shovels, several workers dug 3.6 m into the 8 m deep core wall and found a tree root penetrating the puddle core. They plugged the leak and refilled the excavation in the embankment, with no threat to the overall integrity of the wall. Leaks were not common occurrences, as the earthen embankments were carefully designed. Melbourne's reservoir embankments of earth, rock and concrete have withstood the test of time, although they have been or will be upgraded and modified to ensure their stability and to meet the latest design standards.

Figure 3.3: The Upper Yarra Reservoir was sited in the hilly central highlands where there are spectacular panoramas from the summits (images: D. Lindenmayer and D. Blair).

Box 3.4: A secret wartime tunnel at O'Shannassy Reservoir

During World War II, the Chairman of the MMBW became concerned about what might happen to the records of the MMBW in case of attack. To secure the safety of the records, he proposed a tunnel at a secret location on the top side of the O'Shannassy Aqueduct, downstream from the Commissioners' Quarters.

To ensure secrecy, Engineer for Watersheds, George Skewes, arranged for the day-labour crew from Yan Yean Reservoir to be brought to O'Shannassy to excavate the tunnel, rather than using the local crews based in nearby Warburton. MMBW engineer Roy Kermode, who recounted this story, did not see the secret tunnel and did not know if it was ever used. As far as Kermode knew, the records never left the Head Office in Little Collins Street in the centre of Melbourne. Given the time since the tunnel was excavated, and the secrecy of its location, it is unlikely that it will be found.

Roy Kermode, Engineer for Watersheds from 1962 to 1976, and predecessor to Jim Viggers as Manager Water Supply, Headworks and Distribution (image: Melbourne Water).

government that no timber would be cut from the 18 210 ha that had been reserved for water supply[34] (see Chapter 2). Because of this arrangement, the MMBW had to seek government approval to construct the dam on the Upper Yarra River; approval was granted in 1940.[50] World War II delayed

Figure 3.4: Workers lining up on pay day at the Upper Yarra Reservoir construction site (image: Melbourne Water).

development of the dam works as men, materials and equipment were in short supply during the war. In 1946, work began on laying the first of two pipelines that would be used to carry water from the proposed Upper Yarra Reservoir to the Silvan Reservoir. These pipelines, called Yarra–Silvan Conduits, connected the end of the existing aqueduct at Starvation Creek basin to Silvan Reservoir.[34] The first pipeline was completed in 1953. The work involved tunnelling through Mt Little Joe near Warburton.[5] In addition, a pipeline was laid to connect the O'Shannassy Reservoir to the Yarra–Silvan system to maximise utilisation of the yield of water from the O'Shannassy Catchment.

Design work on the Upper Yarra Reservoir commenced in 1951. Construction on the large dam wall started in 1953 and was completed in 1957.[5,34] A temporary town was constructed downstream from the site of the proposed reservoir, ~27 km from Warburton.[51] Living conditions were improved for the labour workforce – the MMBW had learned from the poor management of the O'Shannassy and Silvan construction camps. The new town for the Upper Yarra workforce comprised buildings instead of tents,

Figure 3.5: a) Construction of the Upper Yarra Reservoir. b) The full reservoir, following completion (image: Public Record Office Victoria).

with three-bedroom houses provided for workers with families and a series of small twin-share huts for single workers. Mess halls, kitchens, bathrooms and laundries, a primary school and entertainment facilities were all provided in buildings instead of tents.[51] Accommodating and feeding ~1200 workers was no small task, and with three shifts a day (8am–4pm, 4pm–midnight, and midnight–8am), the kitchens were staffed almost non-stop to

Box 3.5: From Monte Carlo to the Upper Yarra, via Manhattan – engineering innovation

In 1953, Frank Barnes was given the task of preparing a paper for publication on the hydrology of dam storage using the Upper Yarra Reservoir as a case study. Upper Yarra Reservoir was under design and construction at the time and this provided an opportunity to prove its adequacy to maximise utilisation of the available water from the catchment.

Barnes developed a method of determining the relationship between storage capacity, percentage utilisation of average streamflow (draft) and the probability of the storage running dry or the need to impose restrictions on water use. He developed a statistical method which employed random sampling of a population of streamflows with the same average and distribution as the actual record over 30 years, which was used to generate a simulated record of 1000 years. The model showed that a critical situation at the reservoir would not result from one very dry year but rather from a series of moderately dry years.

The simulated record was used to determine the frequency of large depletions of the water storage at various drafts and sizes. The analysis showed that 200 GL was close to the optimum size for the Upper Yarra Reservoir. The 200 GL reservoir would ensure that 87% of the long-term average streamflow could be drawn from the storage on average each year with only a 1 in 40 chance of the reservoir running dry. Of course, restrictions could be imposed at the appropriate times to reduce the draw-down so that the Upper Yarra Reservoir would never be allowed to run dry. While 87% utilisation is generally regarded as a very high figure, the operation of the dam in conjunction with other storages in Melbourne's system has resulted in a high level of utilisation of Upper Yarra water.

Interestingly, the statistical method developed by Barnes was the Monte Carlo method, but he did not know this at the time. This was the name given to the secret random sampling method developed by nuclear physicists on the Manhattan Project during World War II. Barnes' application of the Monte Carlo method was original and highly innovative. His paper was published in September 1954 and the method was adopted by US authorities and engineers in the UK. It was acknowledged in 1974 in a paper read to the Royal Society in England as the seminal work in using random modelling of river flows.

MMBW design engineer Frank Barnes (image: Melbourne Water).

feed workers before and after their shifts. Workers were paid in cash, which arrived in an armoured truck every second Thursday. Following receipt of their pay packets, workers lined up at the post office branch in the village to deposit their money (Figure 3.4). The regular post truck that took the deposits and other mail to the Warburton post office on Friday mornings often used to carry almost as much cash as the armoured vehicle that had delivered it the previous day, prompting MMBW design engineer Frank Barnes to wonder why no-one ever tried to rob the mail truck.

Upper Yarra Reservoir: revisions to capacity and rate of construction

The construction of the Upper Yarra Reservoir followed consultation with experts from the Bureau of Reclamation in the USA. The dam would be an extremely large embankment, and there were few experts worldwide with the relevant experience to provide advice on its design. The wall itself was an 89 m high earth- and rock-fill embankment. The initial approval in 1946 was for a dam of 130 GL capacity. However, as a result of revised projections

Box 3.6: Too much water – design of the spillway at the Upper Yarra Reservoir

The Upper Yarra Reservoir spillway was designed for a maximum discharge of 2210 cumecs (cubic metres per second). This figure was based on the maximum possible precipitation in 24 hours (as calculated by the Bureau of Meteorology) and was applied to a fully saturated catchment with the reservoir full to the top of the spillway. To give context, 2210 cumecs is 2 210 000 L per second – nearly an Olympic-sized swimming pool running over the spillway every second. Designing a spillway to safely discharge this amount of water was challenging. Frank Barnes calculated that due to changes in direction of the water, it would develop 13 times the normal hydrostatic pressure when passing through the 'ski jump'. Therefore, the ski jump structure had to be designed to withstand this load. The spillway consists of a long side channel overflow to take the water from the reservoir along a chute to an upturned end known as a ski jump (which it resembles). The purpose of the ski jump is to ensure that the discharged water is thrown well clear of the spillway so the huge amount of energy in the water is dissipated safely and does not undermine the bottom of the dam or the spillway.

The Upper Yarra Reservoir spillway (image: Melbourne Water).

for water use and population growth, the capacity was increased to 200 GL in 1951 by raising the height of the dam wall. This change in capacity was shown to be the correct decision, based on modelling by MMBW engineer Frank Barnes, one of a team who worked on the design of the dam wall, outlets, inlets and other associated works (see Box 3.5).

The diesel-powered tractors and machinery used at the Upper Yarra construction site were a progression from the steam-powered earth-moving equipment used at Silvan Reservoir. Construction proceeded at such a rapid rate that the designers had trouble keeping up. Instead of having completed drawings of the proposed work done by draughtsmen, the construction supervisors were working from the pencil drawings of the design engineers. Barnes noted that the pace of construction meant that there was no time for his designs to be checked by anyone else before being issued to the on-site engineers. Naturally uneasy about the prospect of unchecked figures and calculations, Barnes devised a strategy to verify his decisions. He used two different methods to check each calculation and verified that a comparable answer was obtained from both methods. The Upper Yarra Reservoir was completed in 1957 and it filled to capacity in 1959 (Figure 3.5).

Conclusion

Augmentation of Melbourne's water supply was timely given rapid population growth following the end of World War II. Melbourne had a secure water supply with the new 200 GL Upper Yarra Reservoir. The existing challenges to supply may have been met, but with a population surpassing 1 million and with households still wanting lush green gardens, demand for water would continue to increase.

Part 2

1961 TO 2012: A WATER MAN'S PERSPECTIVE

Chapter 4: 1961 to 1984

Introduction

The 1960s and 1970s heralded more activity in the development of Melbourne's water supply system. Droughts featured prominently in shaping the way Melburnians used water, and the source of water for the city. The drought of 1967–68 was responsible for the development of a new scheme in the Thomson River Catchment, 120 km east of Melbourne. The drought of 1982–83 partly contributed to the devastation caused by the Ash Wednesday bushfires in the summer of 1983. In this chapter, we explore how the MMBW tackled the issue of securing the water supply through the development of additional infrastructure. We also discuss a new initiative for the MMBW – the implementation of innovative campaigns to increase public awareness of water usage and reduce the demand for water.

1962: 'The future water supply' report

Melbourne was growing quickly in the early 1960s, with a population of over 1.8 million people.[52] Home-owners were becoming more affluent, which translated to larger houses with more water-using appliances. Extra bathrooms were being built in new houses, and washing machines and dishwashers became increasingly popular. Gardens were still large and the idea of the 'Garden State' encouraged water use.

Albert Ronalds, the then MMBW Engineer-in-Chief, produced the 'Report on the Future Water Supply of the Melbourne Metropolitan Area'[52] in response to increasing concern about how to supply water to Melbourne's growing population. The report included suggestions for water supply options over the following 30 years, based on population projections of 2.5 million people in Melbourne by 1972, 3.4 million by 1982 and 5 million by 2000. The key conclusion of the report was that the existing water supply system had to be amplified to increase the long-term availability and security of the supply of water.

Ronalds recommended the following options:[52]

- augmentation of existing water supplies, including duplication of the Yarra–Silvan Conduit to increase the transfer capacity of water from the Upper Yarra Reservoir, raising the spillway at Maroondah to enable more water to be held in the reservoir, and the construction of the O'Shannassy–Upper Yarra cross-connection to maximise utilisation of water from the O'Shannassy Catchment;
- utilisation of the water from tributaries of the Yarra River upstream of Warburton (Starvation, McMahons and Armstrongs creeks) and the development of the Lower Yarra River scheme (which later became the Sugarloaf Reservoir system);
- reservation of the waters from the Thomson and Aberfeldy rivers;
- diversion of the Big River (a tributary of the Goulburn River upstream of the Eildon Dam, located north of the Great Dividing Range) into the Upper Yarra Reservoir.

Ronalds' recommendation to develop the water resources of the Thomson and Aberfeldy rivers by building a new reservoir revisited an idea first proposed by Ritchie in 1915 (see Chapter 2). The proposal was met with opposition from irrigators in central Gippsland.[5,53] While the MMBW approved Ronalds' recommendations, most of his proposals were large and potentially controversial schemes. Closing catchments that were being logged would be contentious. The report was referred to the state government for consideration; it established a Parliamentary Public Works Committee (the 'Melbourne Metropolitan Future Water Supply Inquiry')[54] to assess Ronalds' proposals.

Analysis of Ronalds' proposals for water supply

Perhaps Ronalds' most (inadvertently) controversial recommendation was to divert the Big River, a tributary of the Goulburn River, into the Upper Yarra Reservoir. Diversion of water destined for the agricultural areas of the Goulburn Valley was opposed by farmers and irrigators, as occurred with the proposed Acheron River diversion in 1902 (see Chapter 2). Rural communities in the Goulburn Valley were against the idea of Melbourne taking 'their' water. The State Rivers and Water Supply Commission also opposed the idea and insisted that there was no spare water for Melbourne in the Big River.[5]

The Parliamentary Public Works Committee inquiry produced progress reports in 1964 and 1966, and a final report in 1967.[54] In the first report, the committee recommended the Yarra River tributaries of Armstrongs,

McMahons, Starvation, Mississippi and Cement creeks be diverted and developed for metropolitan water supply. The committee also suggested that, to protect the integrity of the water supply, land be set aside on the edge of each creek upstream of the diversion weirs (Figure 4.1).[54] However, the committee did not support utilisation of the Big River as it was already committed to the Goulburn Valley irrigation system.[54] The state Premier at the time, Henry Bolte, was philosophically opposed to taking water for Melbourne from north of the Great Dividing Range. He announced before the state election in 1964 that 'not one drop' of water from north of the Great Dividing Range would be used by Melbourne.[5] This was considered a highly political move because it appealed to marginal seats in the farmlands outside Melbourne. Bolte was also under the misapprehension that taking water from north of the Dividing Range was a new idea. It wasn't. Water from Wallaby and Silver creeks, north of the range, had been diverted in 1885 by Davidson and Thwaites and had run into the Yan Yean Reservoir ever since (see Chapter 1).

Final recommendations of the Parliamentary Public Works Committee inquiry

A key finding of the Parliamentary Public Works Committee was that a reservoir should be constructed on Cardinia Creek, south-east of Melbourne, to service the rapidly expanding south-eastern part of the city and to maximise the harvest of water from the Yarra tributaries, and the O'Shannassy and Thomson reservoirs. The committee also recommended that the water from the Lower Yarra River be harvested, via a dam at Yarra Brae on the Yarra River upstream of Warrandyte, with water pumped into reservoirs at Sugarloaf and Watsons Creek near Yering Gorge.[54]

The final report also recommended that the Victorian government authorise the MMBW to develop the resources of the Thomson and Aberfeldy rivers, via the construction of a dam on the Thomson River downstream of its confluence with Talbot Creek. The latter recommendations would involve the catchment areas of the Thomson and Aberfeldy rivers being proclaimed as water supply catchments under the *Soil Conservation and Land Utilisation Act 1958*.[5] The final recommendation for water supply was that the Watsons Creek catchment land be acquired by the MMBW and set aside for future water supply purposes.[54]

Although the Parliamentary Public Works Committee had recommended the development of the Thomson scheme, it was controversial and there was widespread opposition to the plan. Residents and irrigators in central Gippsland were concerned for their water supply from the Thomson and Aberfeldy rivers.[5] The main conclusion of the report was the assessment that the rivers could provide water to Gippsland, Melbourne and the Mornington Peninsula.[54] After the water situation in Melbourne deteriorated during the droughts of the 1960s, and in the face of continued opposition from Gippsland, environmental evaluations were conducted to ensure that the new works would not cause long-term issues for the catchment's industries and residents.[55]

Figure 4.1: The Armstrongs Creek Catchment (image: D. Blair).

The 1967–68 drought

The drought of 1967–68 had a significant effect on Melbourne's water supply, and spurred the development of new water supply options and infrastructure. The Upper Yarra Reservoir fell to its lowest ever storage level. On occasions, the daily evaporation from the reservoir exceeded the daily inflow. The drought conditions led to some small augmentations to water supply in 1967. Cement Creek, a tributary of the Yarra River, was rediverted into the O'Shannassy Aqueduct (having been diverted first in 1939), and new diversions of the Dee River and Walkers Creek at Millgrove were constructed to discharge water into the O'Shannassy Aqueduct.

The catchments of these watercourses were alienated land, so chlorination treatment plants were installed on the aqueduct at Cement Creek and the Dee River to ensure the water quality was not compromised. Additional works to utilise other tributaries of the Yarra River (Starvation, McMahons and Armstrongs creeks), as recommended by the Parliamentary Public Works Committee, commenced in June 1968. The catchments for the tributaries were not closed, nor vested in the MMBW, and their diversion for water supply involved an agreement with the Forests Commission of Victoria that one tributary catchment per year remain open for logging (not contributing to the water supply while being logged). Each catchment would be logged in succession over a period of three years. Because of the logging, each tributary had a chlorination plant installed. In addition, for the first time the MMBW was required to meet specified flows of water on the Yarra River at the Millgrove gauging station by ceasing diversion of water from the Yarra tributaries as necessary. Prior to 1968, the MMBW had no real obligation to release water for environmental flows.

Additional infrastructure and water supply options

The need to maintain supply of water in extremely dry conditions meant that the MMBW implemented all feasible options during the 1967–68 drought. A pumping station and chlorination plant was installed on the Yarra River upstream of Yering Gorge to move water from the river into the Maroondah Aqueduct. However, very little water was harvested from this site as the flows of the Yarra River were extremely low due to the drought. The MMBW also embarked on cloud seeding over the Upper Yarra Catchment. Cloud seeding uses aeroplanes to deploy tiny particles, usually silver iodide, into water vapour (i.e. clouds) to encourage water droplets to aggregate and fall as rain.[56] The trials took place with the assistance of the

Tasmanian Hydro-Electric Commission which had used cloud seeding for many years on the west coast of Tasmania. Several flights were conducted over parts of the Upper Yarra Catchment, but as there were no clouds to seed the project was abandoned.

Water restrictions

The drought of 1967–68 caused significant changes to the way Melbourne used water. It was the first time that an official list of longer-term water restrictions with varying levels of severity was developed and imposed on the population.[57] Prior to this, water restrictions had been imposed only for short-term periods, typically in response to heatwave conditions that put

Box 4.1: From reservoirs to taps – management of the aqueduct system to supply water to Melbourne

Until 1971, no operational procedures existed for the movement of water from catchment reservoirs to service reservoirs and suburban distribution reservoirs. MMBW officer Bill McKee was appointed to work with the watershed managers to write and develop manuals to ensure that their successors would be able to deal with water supply problems in an expanding and increasingly complex system. Previously, operations depended on the small group of men who knew the intricacies of the system. During his time at the MMBW, Jim Viggers developed a very keen sense of how the network operated, and knew how to move water to ensure that supply was kept at an adequate level. The aqueduct system was susceptible to changes in the weather (e.g. storms or cool changes caused an immediate decrease in demand for water), so it was imperative the operational engineers reacted quickly.

From Jim Viggers: 'While I was Duty Engineer at the Preston Reservoir, I was responsible for the water coming on aqueduct systems from Maroondah to Preston, and from Silvan to Olinda. On a very hot night, we'd be flat out trying to keep up supply and maintain pressure in the system, still working up to 11pm. I'd go out and look at the reservoir at the front of the house. I'd have 5 ft of water in the bottom, and we'd need 18 ft by the next morning. I had to trust the decisions I'd made in regulating the supply that the levels would recover by 6am. The next morning, I'd be up at 5am to see the result of the previous night's decisions. We worked like that each day until there was a change in weather. With the Maroondah Aqueduct system, the water required for the next 24 hours at the Preston Reservoir would have to have been released the day before, as there was a 20 hour lag in the water arriving at Preston.'

Times have changed. Pipelines have replaced the aqueducts, providing increased reliability to the supply into service reservoirs. Valves were installed to regulate the amount of water moving through any pipeline. This removed the sense of urgency caused by storms or prolonged hot weather, and increased the availability of water as decisions no longer needed to be made 24 hours in advance.

Box 4.2: Clandestine fishing in water supply reservoirs

Many native and introduced fish species live in Melbourne's water storages but fishing in water supply reservoirs is not permitted. Silvan Reservoir is home to numerous large trout that, on one occasion at least, were caught by workers. In 1973, a construction crew was working on the Yarra–Silvan Conduit inlet to Silvan Reservoir at Ferndale Road. Supply to the conduit had to be shut down and the tunnel under Silvan Monbulk Road drained. Jim Viggers was given the task of draining the tunnel and this revealed large tasty-looking trout.

Jim Viggers recalls, 'We stopped the water flow and were waiting for the tunnel to drain. There were huge trout everywhere! I had two patrolmen with me, and we thought we'd better catch some. We got some chicken wire and sandbags and built a small embankment across the floor of the outlet of the tunnel with chicken wire, allowing the tunnel to drain while holding back the trout. We then went back into the tunnel and caught trout with our bare hands. The construction crew at the other end of the tunnel was also catching fish in this manner. I came out of the tunnel for some reason, just as my boss pulled up in his car. I thought I was in trouble. He looked down and asked what I was doing. I said I was doing a bit of fishing and his reply was 'You know where I live'. He got back in his car and left. At our end of the tunnel we caught more than 20 trout up to 7 kg in weight.'

One of the construction workers had previously had his fishing rod temporarily confiscated by the senior construction engineer after being caught fishing in the reservoir. The engineer just happened to arrive on-site while all the fish were being collected; the worker had caught a nice big trout. The engineer took the rod out of his car and asked if it would be a fair exchange for the trout.

pressure on the metropolitan distribution system. The old aqueduct systems could bring only a certain volume of water to Melbourne per day, and on very hot days demand often outstripped supply. Short-term restrictions (mainly bans on watering gardens) were imposed on a day-by-day basis as necessary. Water engineers assessed the potential demand; if it looked to be greater than the supply, restrictions were recommended and the public were informed via notices in newspapers on the day of the restriction (R. Kermode, pers. comm.). Official water restrictions were also imposed during the 1972–73 and 1982–83 droughts.[57]

New water supply infrastructure for Melbourne

Greenvale Reservoir: western Melbourne

The suburbs of Melbourne had been extending the limits of the city in a south-easterly direction since the beginning of the 20th century, but by the

Figure 4.2: Greenvale Reservoir, completed in 1971 (image: Melbourne Water).

1960s expansion was also occurring to the west and north of the city. More housing in these areas meant that additional water supply services were required. Greenvale Reservoir, an off-stream storage with no catchment, was constructed between 1968 and 1971 (Figure 4.2). It holds 26 GL of water for supply to the developing western suburbs of Melbourne and gives security of supply through balancing seasonal variation in demand.[5]

The site for Greenvale Reservoir had been identified by the MMBW over 50 years earlier when Ritchie recognised it as a potential location for a storage reservoir. He envisaged it would be supplied via an aqueduct and siphon system from a diversion weir on the Watts River upstream of Maroondah Reservoir. However, it is supplied by transferring water from Silvan Reservoir.

Cardinia Reservoir: south-eastern Melbourne

Construction of the dam at Cardinia Creek commenced in 1970, following a period of indecision by the government. A proposal to build the dam had been presented to Cabinet in 1966 but was not approved until late 1967, after the final Parliamentary Public Works Committee report was submitted to the government.[5] The construction period was brought forward to cope with the drought conditions of the late 1960s.

Box 4.3: Jim Viggers – managing Melbourne's water supply

From 1970 to 1976, Jim Viggers was based at Preston Reservoir, the main control centre for Melbourne's water supply. His primary task was to direct the regulation and movement of water through the aqueduct and pipeline systems to the main distribution service reservoirs to meet the day-to-day consumer demand. His position had been created to increase the number of engineers who had skill, knowledge and understanding of the operation of the headworks and regional distribution systems. Until his appointment, only two engineers in the MMBW – the Engineer for Watersheds and the Assistant Engineer for Watersheds – had the necessary experience and skills. They frequently travelled in the same vehicle, which was a concern if they had an accident.

The task of regulating the water supply system was shared, on a seven-day basis, between the Assistant Engineer of Watersheds and Viggers. Both lived on-site at Preston Reservoir and change-over took place on Monday mornings. These two effectively worked 12-day fortnights, so they had every second weekend off. As the duty engineers, they were required to be available 24 hours a day to respond to any issue within the Watersheds Department area of responsibility.

A typical workday would begin at 6am with phone calls from the caretakers of the major metropolitan service reservoirs (Olinda, Mitcham, Surrey Hills, etc.). Viggers would receive information on water levels, flow meter readings, etc., calculate the amount of water that had been consumed in the previous 24 hours and estimate the expected use in the next 24 hours. He would then call the caretakers with instructions on adjustments to the supply system. At 7am he received calls from the caretakers of the major storage reservoirs (Maroondah, Upper Yarra, etc.). Based on the information received, he would calculate daily streamflows and request adjustments, if any, to the amount of water to be released from the storage reservoirs.

The caretakers from the metropolitan reservoirs would call at 11.30am with updates on levels and flows, and call again at 4pm. The consumption for the 10 hours since 6am would be calculated and the system again adjusted, if required, to satisfy expected demand for the next 14 hours and to achieve optimal operational levels in the service reservoirs by 6am the next day.

On very hot days when water usage was high, caretakers would often be required to call back with reservoir water levels at 8pm or even later to enable further adjustment to the system, if necessary, to achieve recovery in levels.

After his promotion to the position of Deputy Watersheds Manager in 1976, Viggers assumed responsibility for the operation and maintenance of the headworks and regional distribution systems. He also worked to ease the work conditions of the Duty Engineer. He established a system where operators were appointed to complete many of the routine tasks and he increased the number of engineers to six. Hence, the frequency with which a person was rostered on as a Duty Engineer was reduced, and working hours became more flexible.

Box 4.4: Unwelcome additives to water – turbidity, seagulls and jellyfish

Turbidity at Greenvale

Vegetation clearing of the land that would be inundated to form Greenvale Reservoir exposed an area of decomposed granite on the eastern shoreline below the reservoir's high-water mark. This material was quite hard and stable until it was exposed to water and wave action. On initial filling of the reservoir, the decomposed granite began to erode, releasing fine particles of clay that stayed in suspension in the water. The result was turbid (murky) water, which gave the reservoir the look of a farm dam. This was highly undesirable from a water supply perspective as the water was obviously unfit to drink.

Alum was suggested as a potential remedy, as it is commonly used to flocculate particles out of suspension in water, but attempting to mix it into a reservoir holding ~16 GL of water was quite a task. An attempt was made, however, using a tanker to add 30 000 L of alum to water entering the reservoir at the inlet from Silvan Reservoir. This would allow the natural churning of the incoming water to mix and help distribute the alum. It was a partial success – the water became clear, but only in a 20 m radius of the inlet.

Another suggestion was to use fingerling trout. The theory was that they would filter the clay particles with their gills as they breathed the oxygen from the water. An order was placed at a trout farm near Ballarat, and thousands of baby trout were released into the reservoir. However, gills work to remove oxygen from the water so the fish can breathe – gills are not all-purpose filters. Perhaps unsurprisingly, the water did not clear after trout were added to the reservoir. The trout were never seen again, although Viggers knew caretakers who saw large trout in the reservoir years later. Finally, it was decided to drain the reservoir to expose the offending area and to stabilise it with rock beaching to prevent further erosion. Water quality was restored, and the reservoir was filled and brought into service.

Seagulls at Somerton

In 1973, a plague of crickets descended on parts of Melbourne's northern suburbs. Seagulls and other birds took advantage of the situation, and after eating their fill congregated in the water and along the banks of the Somerton Service Reservoir. Somerton was a trapezoidal reservoir with sloping bituminised sides. The birds posed a potential problem to the quality of the water supply, and weren't welcome in large numbers. The Engineer-in-Charge decided that the birds must be scared away to prevent pollution of the water, and procured a gas-fired scare gun. The birds flew into the air with every blast, but kept coming back. The unforeseen problem was that the frightened birds defecated as they took off, which rendered the sides of the reservoir white as they were repeatedly scared away. As this approach proved ineffective the reservoir was withdrawn from service until the crickets and birds departed.

Jellyfish at Yan Yean

Also in 1973, there was an outbreak of small (2–3 cm in diameter) freshwater jellyfish at the Yan Yean Reservoir. They were not dangerous, but they were crushed in the mesh screens of the outlet works, resulting in tiny pieces of jellyfish entering the outlet main and potentially being piped to homes and appearing in people's drinking water. The problem was solved by reversing the flow in the outlet main; this method of operation continued until the outbreak of jellyfish disappeared.

Figure 4.3: Cardinia Reservoir, completed in 1974 (image: Melbourne Water).

Cardinia Reservoir is large, with a 1.5 km long earth- and rock-fill embankment. With a capacity of 283 GL, it was the MMBW's largest water storage facility and it was believed that it would 'drought-proof' Melbourne. As an off-stream reservoir, Cardinia would store the excess winter flows from the Upper Yarra Reservoir, Yarra tributaries and the O'Shannassy and

Box 4.5: Finding the way – preliminary surveying for the Thomson River scheme

When the Thomson River scheme was first confirmed as a water supply option in the mid-1960s, detailed maps were needed before work could begin. The only maps of the area that existed at the time were old military maps, which, at a scale of one mile to one inch, did not provide the level of detail required for engineering works. In 1967, a surveying team led by MMBW surveyor Russell Read was dispatched to set up targets for aerial photography of the Upper Thomson area. The bush was wild. The few tracks that had existed before the 1939 bushfires were overgrown and needed to be cleared. The aerial photography team also needed large clearings with big white crosses set out on the ground – so they could aim their cameras at the right places. Read's team spent two months moving through the steep and densely forested terrain co-ordinating the control points for the aerial team, cutting down trees to make clearings and setting up new tracks for vehicles.

Box 4.6: Can you see me? Surveying in dense bushland

Surveying teams worked in tough conditions when conducting initial surveys of the sites for the Thomson River scheme. Often forestry crews were brought in to clear trails for the surveyors. MMBW surveyor Russell Read found a novel way to help the foresters see the lines they had to clear, by using weather balloons. He obtained large weather balloons (approx. 1.8 m diameter) from the Bureau of Meteorology and filled them with hydrogen – it was cheaper than helium, although far more dangerous. The aim was to let the balloon up into the air above the tree tops at a designated point, to show the foresters a straight line along which they would clear trees. It was not as easy as hoped, as the very large balloon, when inflated, was difficult to manoeuvre through the scrub to the next point. The system was altered after a worker, towing along an inflated balloon, fell into a disused mineshaft. The survey team heard a yell from the (uninjured) worker and an enormous bang – the balloon, too large to fit down the hole, had burst. Following this incident, Read needed to think of a better way to manage the balloon. The idea was too good to abandon completely – the balloon above the trees gave the foresters a very clear line to work towards – but the large balloons were too cumbersome. So he went to the local newsagent and bought a supply of coloured party balloons. Read filled 20–30 balloons with hydrogen, tied them in bunches and let them up into the canopy instead. They must have been quite a sight in the middle of a Mountain Ash forest.

Thomson reservoirs. Water from these systems was transferred to Cardinia via Silvan Reservoir.[5] Cardinia Reservoir was sited at a strategic location that enabled it to become a water supply resource for the south-eastern suburbs of Melbourne as well as the Mornington Peninsula (Figure 4.3).

Commencement of the Thomson River scheme

The first diversion of water from the Thomson Catchment began during the 1967–68 drought. This was an open-earth channel constructed between the Upper Thomson River and 18 Mile Creek (a tributary of the Upper Yarra River) to capture more water for the Upper Yarra Reservoir. Prior to any diversions being constructed, MMBW surveying teams had to decide the best location for the diversion and earthen channel across a large ridge that separated the two watercourses. The team measured altitude by barometric heights; although this method had a degree of uncertainty, it was the best available at the time. The altitude measurements were necessary to plot the flattest course for the channel. Bulldozers and construction machinery were then driven into the site but fuel had to be delivered by helicopter due to limited road and track access, especially in winter. Another diversion weir

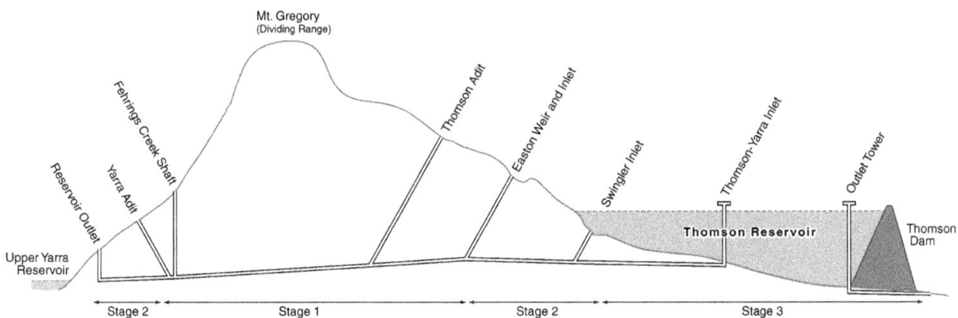

Figure 4.4: Schematic diagram showing the stages of the Thomson scheme, with tunnels and construction of the dam wall (image: C. Hilliker).

was constructed in 1971 to divert water from the West Tanjil River to the Upper Thomson River, where water would be captured by the diversion to 18 Mile Creek.

The scheme for the development of the Thomson River was segregated into three stages, comprising two tunnelling stages and a final combined tunnelling and dam-building stage. The entire project took 16 years to complete.

Stage 1: tunnelling to Upper Yarra Reservoir

Stage 1 of the Thomson River development was a 20 km tunnel which commenced in 1973. It ran east–west between a new diversion on the Thomson River, called Easton Weir, and Fehring Creek, a tributary of the Yarra River upstream of the Upper Yarra Reservoir (Figure 4.4). The Easton Weir would divert water from the Thomson River into the tunnel once construction of this section of the tunnel was complete. At the other end of the tunnel, water flowed from a shaft into Fehring Creek and then into the Yarra River, destined for the Upper Yarra Reservoir.[55]

Construction began with three access tunnels (adits) being excavated: at the eastern end (Thomson adit), Easton Weir (Easton adit) and the western end (Yarra adit). The plan was to use a large rock-boring machine to excavate the tunnel in one direction from the Yarra adit to the proposed Swingler Weir on the Thomson River downstream of its confluence with the Jordan River. This was unsuccessful because the work met seams of intrusive quartz rock which ground away the cutting heads of the boring machine, causing progress to slow significantly. Eventually, the only way to make the construction deadline was to use traditional drill and blast methods on three faces – from the Yarra adit eastwards, and from the Thomson adit in both west and east directions (Figure 4.5).

Figure 4.5: a) Thomson tunnel, Thomson–Yarra Junction. The tunnel is 3.6 m high and in fresh rock. The tunnel leading to the Easton Junction is on the right. The adit tunnel is to the left. The two vent lines were 900 mm diameter and extracted stale air from the tunnel faces. The locomotive was diesel-powered with chemical scrubbers to remove poisonous fumes. b) Yarra Portal. The tunnel entry (adit) is to the left of the bridge under the white posts. Rock was transported out of the tunnel in rail cars and across the river to spoil dumps to the right of the photo. The Yarra River is in the centre, the generator building is in the background and workshops are lower left (images: Austin Byrne).

Box 4.7: Breaking through to the other side

The breakthrough between the faces of the Thomson tunnel Stage 1 was tense, as no-one was sure that the two ends of the tunnel were actually going to meet in the middle. The headings and bearings all corresponded, but doubt would be removed only with the actual breakthrough. The surveyor on site, Graeme Barber, inspected the preparations for the breakthrough from the Upper Yarra side. The tunnelling team set up an exploratory bore to drill through the rock to the other side. This bore was positioned ~180 cm above the floor, at the approximate mid-point of the tunnel height. Barber left the team to it, and set out to view the breakthrough from the other side. He drove more than 30 km to the Thomson end and, with the tunnelling team on that side, waited for the bore to work through the rock. They listened to rock being ground, and heard the crack as the final fragments of rock gave way. The bore came through the rock ~60 cm off the tunnel floor at the Thomson end – which was highly concerning given the bore had been set up at 180 cm above the floor at the Upper Yarra end. Concerned that the tunnels wouldn't match up, Barber jumped in his car and returned to the Upper Yarra end, where he found that the tunnelling team had had to reposition the bore 60 cm above the ground because it was too unstable at the original 180 cm position. The final variation at the meeting point of the two tunnels was only ~5 cm.

Tunnelling from the Yarra to Easton and Easton to Yarra required very careful surveying to ensure the tunnels would meet. The surveying and construction teams were concerned about potential refraction of the line of sight of the infrared surveying equipment within the tunnels. Refraction could give inaccurate readings, which would lead to a potential mismatch of the heights of the different headings of the tunnel. The surveyors developed a zigzag method, measuring first from one side of the tunnel then alternating to the other side for the next reading, to ensure they were not receiving inaccurate readings caused by refraction (R. Read, pers. comm.).

Stage 2: extension of the tunnel

The second phase of the Thomson–Upper Yarra tunnel was a 5.1 km extension at the western end from the Yarra adit to a new shaft upstream of the Upper Yarra Reservoir and a 5.6 km extension from the site of the proposed Swingler Weir to the Easton adit (Figure 4.6). The Stage 1 tunnel was plugged downstream of the Fehring Creek shaft and upstream of the Easton adit. This enabled the tunnelling work to proceed and at the same time allowed water to be transferred from the Thomson River via the Easton Weir to the Upper Yarra Reservoir.[55] Following completion of Stage 2 works, the Fehring Creek shaft was filled with rock.

Figure 4.6: Swingler Weir, later submerged after the Thomson tunnel and reservoir were completed (image: Melbourne Water).

Stage 3: final extension of the tunnel and construction of the Thomson dam wall

Stage 3 commenced in 1976 and comprised a 6 km extension of the tunnel from Swingler Weir to Bells Portal located on the Thomson Reservoir upstream of the wall. Work on the earth- and rock-fill dam wall commenced in 1981 and was completed in 1983 (Figure 4.7).[53,58] On completion, and when the reservoir was full, the inundated area would extend 23 km upstream from the embankment of the dam and store in excess of 1000 GL, making the Thomson Reservoir the largest storage in the MMBW system. However, the annual water yield of the catchment was ~230 GL and the difference between yield and capacity meant that the reservoir would take many years to fill, even if no water were drawn from it. Indeed, the Thomson Reservoir took seven years to fill after the completion and reached full capacity for the first time in 1990. Hence, the Thomson Reservoir requires careful management as poor decisions on water usage can have long-term effects.

Fluoridation of Melbourne's water

The state government decided in 1972 to include fluoride in the water supply to help protect the teeth of Melburnians (legislated by the *Health*

(Fluoridation) Act 1973). The MMBW was directed to install fluoridation plants as necessary to ensure that all water in the reticulated supply was fluoridated; the cost of the plants was met by the government. Plants were installed on the outlet mains at Silvan, Cardinia, Greenvale and Yan Yean reservoirs, on the Winneke–Preston main at Research, at the supply offtake to Whittlesea township on the Clearwater Channel, and on the Monbulk/ Dandenong Ridge supply system from Silvan in 1973. With the abandonment of the Maroondah Aqueduct downstream of Yering Gorge, the fluoridation plant at Research was decommissioned and a new plant installed at Winneke Treatment Plant located at Sugarloaf Reservoir.

Sugarloaf Reservoir and Winneke Treatment Plant: construction and controversy

September 1972 was the driest month for a century, and the dry weather continued into 1973.[5] The drought highlighted shortcomings within the supply system that could cause critical water shortages while large projects such as the Cardinia Reservoir and the Thomson River scheme were still being constructed.[55] The Parliamentary Public Works Committee had recommended that a reservoir be built on the Yarra River at Yarra Brae, near

Figure 4.7: Construction of the Thomson dam wall (image: Melbourne Water).

Box 4.8: Integrity of the Thomson outlet tower – infrastructure challenges

Defects in the concrete footing of the upper outlet tower of the Thomson Reservoir were discovered when laboratory analysis of a sample batch of concrete showed that its strength and integrity were significantly below minimum requirements. The batch tested had a strength of only 2 megapascals (MPa), whereas concrete for industrial purposes usually has a strength of 28–70 MPa depending on the application.[64] The required strength was 40 MPa. Concrete that can withstand only 2 MPa would not even support a person standing on it. The engineers were perplexed until they found that the 'unders and overs' control of the cement hopper at the batching plant had been switched off. Cement is powdered and does not run like sand. Instead, it tended to land in large clumps on the scales in the batching plant. Big clumps landing on the measuring plate made the needle on the scales fluctuate wildly and the 'unders and overs' control was to prevent the batch moving until the needle stopped. By overriding the control, batches of concrete moved on quickly before the needle had settled. This resulted in varying amounts of cement in each batch. Given that cement is the glue that holds concrete together, this was a major issue. Rob Cranston, Deputy Resident Engineer, was responsible for identifying the problem and rectifying the situation. This included identifying all the batches that had been made while the control was switched off, using coring and forensic methods to find where those batches had been placed, and removing the defective concrete and replacing it with properly measured batches. The volume of concrete involved was considerable, exceeding 200 m^3.

Construction of the outlet towers at the Thomson Reservoir (image: Melbourne Water).

Box 4.9: Rawson – a new town for the Thomson Reservoir construction

Over 700 workers and 200 supervisory staff were employed on the construction of the Thomson dam wall and Stage 3 tunnel. Workers were normally housed in demountable or temporary accommodation. However, the large number of people to be accommodated for this project meant an entire town was needed. In 1975, the Victorian Parliament appointed a committee to find a suitable place for a township close to the Thomson dam site. Land from three farms was purchased by the MMBW at the chosen site, and the town of Rawson was 'born' in 1977.[65] As work on the project wound down and workers left, the houses and facilities were progressively made available to the public for purchase. The population of Rawson was 325 in 2012.[66]

Warrandyte.[54] It also recommended a potential water supply storage option at Christmas Hills (the site of Sugarloaf Reservoir) and the development of Watsons Creek as a water storage.[54] The Lower Yarra area had been identified as a potential source of water in the Thwaites report of 1901, and a water gauging station had been installed on the river at Warrandyte in the late 1800s.

Environmental impact study

In 1973, the state government decided that the Lower Yarra scheme could proceed,[59] despite widespread public opposition to the project in the Warrandyte and Bend of Islands areas over the position of the potential reservoirs. The public reaction to the Yarra Brae aspect of the proposal meant that, for the first time in Victoria, an environmental impact study was conducted at the proposed construction sites.[59] It examined the long-term environmental implications of the three parts of the larger Yarra Brae–Sugarloaf–Watsons Creek proposal in terms of inundation effects and construction and operational aspects, and the effects on vegetation, wildlife, habitats and human activities within the proposed construction areas.

Following consultation and evaluation, the construction of the Yering Gorge pumping station, the Sugarloaf Reservoir and Winneke Treatment Plant was adopted by the MMBW as the best option, with the Watsons Creek storage proposal deferred for possible future development.[59] The Yarra Brae storage option on the Yarra River was found to be inappropriate, as it would inundate sensitive environmental areas.

Box 4.10: The floating pipe from Sugarloaf to Preston

In the early 1970s, as part of the Lower Yarra scheme, work commenced on laying the 2100 mm Winneke–Preston main. This would replace the Maroondah Aqueduct, and supply water from the Winneke Treatment Plant to Preston Reservoir. As part of the works, the pipeline had to be laid through a 1.6 km tunnel.

The traditional method of laying pipes within tunnels was to excavate the walls and floor in order to accommodate both the proposed pipe and a rail track. External to the tunnel, lengths of pipe were laid progressively onto rail trolleys then the pipeline was rolled into the tunnel, like a train. On completion, the track and trolleys were not recoverable. The void or air space external to the pipe in the tunnel was filled by holes drilled from the surface above the tunnel, through which cement grout was pumped.

A unique method for laying the pipes in the 1.6 km tunnel was developed by MMBW engineer Rob Cranston. Cranston's ingenious idea was based on the Archimedes principle and entailed using water to support the pipe and float it along the tunnel. This approach meant minimal excavation of the tunnel, significantly less than required by traditional methods. Using Cranston's method, the only work would be levelling the floor of the tunnel and providing 50–100 mm of clearance for the floating pipeline. Prior to laying the pipe, cradles were placed on the floor along the length of the tunnel. These were required to support the pipeline as it settled into its final position when the water was drained from the tunnel. Rail tracks and trolleys would not need to be laid and left in the tunnel after completion of the works. Both these features would result in significant savings in overall costs.

The first challenge was to keep the water in the tunnel so the pipe sections could float. The team built a weir which comprised three rubber flaps at the downstream entrance to the tunnel. This weir retained water at the level needed to float the pipe. It also created a simple seal as the pipe was winched into the tunnel. The leading section of 2100 mm pipe had a half-hemispherical bulkhead welded to it to prevent the water from entering the pipeline (thereby ensuring it would float). Guide wheels were attached at the top and springing lines of the pipe. Rail track was placed inside the bottom of the pipe to add weight, thereby acting as a keel to stop the floating pipe from rotating. The rail track would be used after laying the pipe to facilitate access for the fixing of spud bars via predrilled holes in the pipes and the grouting of the void between the pipeline and the tunnel.

At the downstream end, rail track was laid external to the tunnel. Rail trolleys with pipe cradles supported by hydraulic jacks were used to assemble and weld the joints of up to 10 lengths of pipe at a time. Each pipe was 12 m in length. The welded pipe was then winched into the tunnel. As the trolleys supporting the pipe reached the sealed weir, the jacks on the trolleys were lowered to allow the pipe to move over the weir and onto the water. The lengths of pipe were then floated into the tunnel. This process was repeated until the bulkhead reached the east end of the 1.6 km tunnel. Then the water was drained, lowering the pipeline onto its cradles before the final fixing and grouting. The rail track in the bottom of the pipeline was recovered and the pipeline brought into service.

The innovative contraption comprising the 1.6 km of floating pipe was called the 'HMBS *Cranston*' ('Her Majesty's Board Ship *Cranston*').

The floating pipeline, HMBS *Cranston* (Image: R. Cranston).

The construction of Sugarloaf Reservoir

Sugarloaf Reservoir is an off-stream storage, with a pumping station located at Yering Gorge to harvest water from the Yarra River. The proximity of the pumping station to the Maroondah Aqueduct meant that water from this source could also be pumped into Sugarloaf Reservoir. This was first time

Box 4.11: Turbidity or no turbidity? Water quality decisions

Chlorination plants have been used since the 1940s to treat drinking water that had moved along open-channel aqueduct systems from major reservoirs to the metropolitan service reservoirs. Occasionally, water became discoloured due to debris and soil washing into the aqueducts during storms or bad weather. The MMBW adhered strictly to the guidelines for acceptable water quality and managed microbial and chemical content in the water, as well as the physical properties of the water such as turbidity. Small amounts of turbidity are within the safe range for consumption, even though turbid water may not look attractive.

In the 1970s at the Preston Reservoir office, Jim Viggers and his colleagues had a rather unorthodox (by today's standards) method to test whether water was too turbid for consumers: 'After heavy rain, the water would be discoloured by the time it arrived at Preston Reservoir from Maroondah Aqueduct. We would fill up a glass milk bottle with water from upstream of the Preston Reservoir. If we could see through it and the water appeared clear, supply from the source was continued. If the water lacked clarity, it was considered unfit for consumption and discharged to waste. Now it's just not acceptable if it's cloudy at all.'

that Melbourne would be harvesting water from an inhabited area,[5] meaning that full treatment of the water (via the Winneke Treatment Plant) would be required to meet potable water standards.

Construction began in 1976 on the 92 GL Sugarloaf Reservoir, including an earth- and rock-fill dam wall, the pumping station at Yering Gorge on the Yarra River, and the Winneke Treatment Plant. The project was completed in 1980. Sugarloaf Reservoir played a critical role in supplying the system until the Thomson Reservoir became operational. Indeed, the extra water from the Sugarloaf Reservoir meant that water could be retained in the Thomson Reservoir, enabling it to fill to capacity by 1990.

Additional options to secure Melbourne's water supply

Watsons Creek and the Aberfeldy River

Two other potential water supply options recommended by the Parliamentary Public Works Committee included an off-stream storage at Watsons Creek, a tributary of the Lower Yarra River, and utilisation of the Aberfeldy River, a tributary of the Thomson River. The Watson's Creek proposal was deferred in 1975, as the construction of the Sugarloaf and Thomson systems postponed the need to build additional infrastructure until at least the 1990s.[5]

Box 4.12: The Aberfeldy River – a missed opportunity?

One of the MMBW's core responsibilities was the development of strategic water plans to ensure that supply kept pace with forecast demands. Several options outlined by the MMBW during previous strategies were not pursued, for various reasons, and with the completion of the desalination plant (see Chapter 5) it is unlikely that they will be implemented in the foreseeable future.

One example is the Aberfeldy River, a tributary of the Thomson River. With an annual stream flow of 44 GL, it was first suggested as an option for water supply in 1962, as an integral addition to the Thomson River scheme.[54] At relatively low cost, a weir could be constructed on the Aberfeldy River to divert water via a tunnel into the Thomson Reservoir.

In 1967 the Parliamentary Public Works Committee recommended that the Aberfeldy River Catchment be reserved for metropolitan water supply.[54] However, in 1975, construction was deferred following an environmental study where the yield of the Aberfeldy River was considered to be too low to be useful for water supply based on revised population growth projections.[55] Despite the deferral, the government recommended that the Aberfeldy River continue to be represented in plans as a future option for augmentation.[60] Water use and population growth projections in 1982 indicated that major headworks augmentation would be necessary by the mid-1990s and the Aberfeldy scheme would be one of the options. However, in 1986 the yield of the Aberfeldy River was again considered too low to be useful in the Thomson system, and the proposal was not recommended for further consideration.[60,61] Based on population forecasts and water usage patterns in 1991, the need to augment the water supply system was deferred until 2006.

The original Thomson River development, as recommended by the committee, included a diversion weir on the Aberfeldy River downstream of the river's confluence with Donnellys Creek, with a tunnel to the Thomson Reservoir.[54,60] Despite numerous recommendations that the Aberfeldy River could be utilised quickly and cheaply for water supply, and warnings that other existing water supply infrastructure would need substantial augmentation by the mid-1990s,[54,56,60,61] the Aberfeldy River diversion has never been implemented (see Chapter 5).

'Planning for an uncertain future' report (1981)

Twenty years after Ronalds' long-term water supply strategy,[52] the MMBW prepared a new water supply strategy. Its focus was on maintaining adequate supply for the increasing demand for water. Key elements included public education programs on conserving water (the highly successful 'Don't be a Wally with Water' was part of the campaign), and increasing the efficiency

Box 4.13: On the front line – Viggers' evaluation of water supply options during the 1980s

Following the completion of the Thomson Reservoir in 1983, Jim Viggers was looking ahead to identify potential sources of water for the future. He was concerned that prolonged periods of drought would result in the Thomson Reservoir being unable to be filled, reducing the security of water supply to Melbourne. He revisited Ronalds' proposal that water could be diverted from either the Big River, sending water to the Upper Yarra Reservoir, or from the Black River, sending water to the Thomson Reservoir. Both these options had the advantage of using gravity to feed the water from the diversions, reducing costs in the long term.

The rationale for making these suggestions immediately after the completion of the Thomson Reservoir was that Viggers saw 2004 as the time when crucial augmentation of the water supply would be required due to supply capacity and the expansion of Melbourne. He suggested to the MMBW General Manager it would take 10 years to win political agreement for approval of the schemes. Therefore, preparatory work needed to start by the late 1980s. While his suggestions were not taken any further by the MMBW at the time, he was correct about the timing of the next critical period for Melbourne's water supply.

of water use by implementing new technology such as dual-flush toilet cisterns.[60]

Recognising that the whole of Melbourne's water supply was based on streamflows from water catchments, the report concluded that the existing water supply resources would be exhausted by 2000, based on projections of population and demand.[60] It listed several other options for streamflow-based supply, including several from north of the Divide – the Big, Black, Yea and Acheron rivers, and the Macalister River in Gippsland[60] – and explored other mechanisms for obtaining drinking water. Recycling of wastewater, exploiting groundwater, domestic rainwater tanks and desalination were all suggested as potential options for water supply, and were revisited in later water supply strategies.[56,60–62]

Bushfires in the catchments: the Ash Wednesday fires of February 1983

Bushfires during November 1982 damaged the catchment at Wallaby Creek, and MMBW fire fighters were mobilised to fight that fire. In February 1983, large parts of Victoria, including water catchment areas, were damaged by bushfires. The 1983 Ash Wednesday fires began on a day of extreme heat and high winds, and burned part of the O'Shannassy Catchment, the Yarra tributaries and Upper Yarra catchments. In the Upper Yarra, extensive areas of catchment were burned, from south of the dam wall east almost to the location where the Upper Yarra River flows into the reservoir. The MMBW, in conjunction with the Country Fire Authority and the Forests Commission of Victoria, worked for a week to bring the blazes in the O'Shannassy and Upper Yarra catchments under control.[63] Workers built firebreaks and backburned. These actions, combined with a weather change bringing widespread rainfall, prevented more of the Upper Yarra Catchment from burning. In total, over 42 000 ha of catchment were burned.[63]

Conclusion

The period from the 1960s to mid-1980s involved key decisions on the future of Melbourne's water supply. The development and construction of the largest reservoir in the water supply system, the Thomson Reservoir, gave the MMBW some breathing space in terms of meeting demand for water usage. However, the large capacity of the Thomson Reservoir was offset by the prolonged time require to fill it to capacity. Several attempts were made by the MMBW to have the government consider various options to increase water supply, given the common view that the next steps for securing water supply would be needed by the 1990s.

Chapter 5: 1985 to 2012

Introduction

The mid-1980s to the mid-1990s was a relatively quiet period for the development of Melbourne's water supply. Following the completion of the Thomson Reservoir in 1983, the water supply was secure for the foreseeable future, as long as rain continued to fill the new reservoir. The Thomson Reservoir filled to capacity for the first time in 1990. The structure of the water industry was rationalised by the government, which led to the formation of Melbourne Water Corporation in the early 1990s. An unprecedented drought from 1997–2009 stimulated the first new water supply projects in over two decades – the North-South Pipeline and the desalination plant. Both projects are discussed in this chapter.

Water supply strategies: 1986 and 1991

Managing demand

Historically, Melbourne had a high per capita usage of water. In a submission to the 1889 Royal Commission on the sanitary condition of Melbourne (see Chapter 1), William Davidson recognised that water usage in Melbourne at that time (estimated to be ~340 L per person per day) was far higher than that of Great Britain, where the highest usage was ~227 L per person per day.[14] Davidson attributed the high demand to Melbourne's hot climate, which he suggested led to people bathing more frequently and watering their gardens more enthusiastically.[14] The pattern of high water usage continued throughout most of the 20th century.

Prior to the 1982–83 drought, Melbourne's water usage grew at a rate of 3% per year, which meant that the existing water supply system would have to be doubled in capacity every 20 years.[56] However, demand management campaigns proposed in the 1982 water supply strategy (see Chapter 4)[60] were implemented; these began to reduce overall water consumption in Melbourne by the mid-1980s.[61] Continued development of demand management campaigns was endorsed by the authors of the MMBW's water supply strategy in 1986.[61]

By 1991, the combination of the demand management campaigns and the introduction of a pay-for-use water billing system resulted in a 16% reduction in the overall average yearly water consumption, relative to levels in the early 1980s. This included a 26% reduction in water use over the summer peak period.[56]

Philosophical changes in attitudes to water use from the 1980s meant that when Jim Viggers retired from the MMBW in 1990, things were vastly different from when he started in the 1950s. Early in his career, the MMBW had actively sought ways to sell water to generate income, especially when storage reservoirs were full or close to capacity. He had managed a system that was stretched to its limits during peak times in summer, with the aqueduct system requiring almost constant regulation to ensure water would be delivered to consumers in the city (see Boxes 4.1, 4.3). Less water was used annually by Melbourne in 2012 than when Viggers retired in 1990, even though the population and the area supplied were significantly larger (2.6 million in 1990 v. 4.4 million in 2011). In 2012 Melburnians used only ~1.8 GL during peak days, whereas Viggers had to deal with peak days requiring in excess of 3 GL.

Reduced per capita demands on a day-by-day basis, together with the filling of the Thomson Reservoir, perhaps gave Melbourne Water and the state government a false sense of security about the long-term adequacy of the infrastructure that supplied water to Melbourne.

Options for new infrastructure and water supply

Projections for water usage had highlighted the importance of commencing new infrastructure projects by the mid-1990s to meet predicted demands. Strategies to utilise the Aberfeldy, Big, Black, Macalister and Yea rivers had been suggested repeatedly[52,60,61] (see Chapter 4) and continued to be proposed until 1991. However, no augmentations to the water supply system were implemented from those sources.[56] The attitude of Melbourne Water and the state government seemed to be that augmentation of the system was not necessary, and it appeared that little consideration was given to the lead-in time required for large infrastructure projects (see Box 4.13).

Options such as recycling water, utilising groundwater, installing rainwater tanks, building desalination plants and harvesting water from icebergs were repeatedly proposed as potential alternative sources of water. By 1991,

Box 5.1: A cool drink – icebergs for water supply

Although the option of using icebergs to supply water was posed in the 1980s and 1990s, the suggestion was not novel. Estimates of the cost for towing icebergs from Antarctica to other cities, such as Adelaide and Perth, had been made by other organisations. The MMBW extrapolated the results of these previous assessments to estimate the cost of doing the same for Melbourne. The potential freshwater yield of a large iceberg is substantial – a 200 GL iceberg (~250 m in height) would supply Melbourne for ~6 months (at 1991 levels of water use). However, the iceberg would have to be moored 50–100 km offshore at Cape Otway or the coast of Sale. Bass Strait is too shallow to accommodate an iceberg with an estimated 200 m draught.[56] The cost of mooring would likely be prohibitive, without even considering the logistics and cost of installing mechanisms and pipes to collect water from the iceberg and send it ashore.

recycled water was seen to be a reasonable option in terms of cost/benefit, and water from South-Eastern Purification Plant was being reused for irrigation in the Cranbourne area.[56] However, the uses of recycled water are limited as it is not treated to potable standards. It was also considered unnecessary to reuse treated water during periods of 'normal' weather conditions when there were average rates of streamflow and catchment yield.[56]

Rainwater tanks in areas of high rainfall have the potential to augment the supply to houses, but the variable quality of tank water and the small capacity of tanks limit their usefulness for widespread implementation.[56] Moreover, the installation of rainwater tanks was discouraged during the 1970s and 1980s due to concerns over water quality.

Another option that did not gain traction was harnessing icebergs as a supply of fresh water (see Box 5.1). Acquiring, towing, storing and converting the ice to potable water seemed unfeasible.[60] Desalination was considered cost-prohibitive,[56] although technology for desalination advanced rapidly during the 1990s and 2000s (see 'Utilising the new infrastructure' below).

Cloud seeding trials in the Thomson Catchment

A further option for augmenting Melbourne's water supply was cloud seeding, as had been done in Tasmania. Cloud seeding over MMBW water catchments, using silver iodide, was first attempted during the drought of 1967–68 but was unsuccessful (see Chapter 4). From 1987, in conjunction with the CSIRO, the MMBW conducted a six-year trial to investigate the potential of cloud seeding to increase catchment rainfall and streamflow.[56]

The Thomson Catchment was selected for the trials because of the reservoir's large capacity. For precipitation to occur over the Thomson Catchment, suitable clouds had to be seeded upwind of the catchment. The seeding team flew in very difficult conditions – stormy weather offered the best conditions for cloud seeding (see Figure 5.1).

The results of the trials were inconclusive. There was no evidence that cloud seeding had increased the water yield to the Thomson Reservoir. One of the primary differences between the sites being seeded in Tasmania and those in the Thomson Catchment was that the west coast of Tasmania was a much larger target area for rainfall. In contrast, the Thomson Catchment was a long, narrow strip of land running north to south in the rain shadow of the Great Dividing Range. Cloud seeding was not attempted again by the MMBW.

Climate change

The 1982, 1986 and 1991 water supply strategies[56,60,61] noted that climate change would be a factor to consider in the management of Melbourne's water supply and it was likely to have a negative influence on water cycles and streamflows. The 1986 water strategy recommended that the CSIRO and MMBW collaborate to develop a detailed understanding of how

Figure 5.1: Cloud seeding – workers check the level of silver iodide being released over clouds. The person on the right is Bob Dorrat, who was the project manager of the cloud seeding project (image: Melbourne Water).

catchments may be affected by climate change.[61] Twenty years after that recommendation, the CSIRO and Melbourne Water released a joint report on the implications of climate change for Melbourne's water supply.[67]

Increased summer average and yearly average temperatures, reduced rainfall, and more variable and extreme weather events were predicted to occur in the Melbourne region as a result of climate change.[67] The potential implications for the catchment areas included an increased risk of bushfires, resulting in the prevalence of young (high water-consuming) regenerating forest (see 'The water catchment forests' below), decreased streamflow due to reduced rainfall, and a decline in water quality due to changes in environmental conditions caused by fire and dry weather.[67]

Like the majority of cities worldwide, Melbourne was entirely reliant on large bodies of water fed by rivers and streams, making it vulnerable to changes in climate. A key recommendation in the CSIRO/Melbourne Water climate change report was to investigate alternate means of water supply, such as desalination or groundwater, to secure future water supply.[67]

The water catchment forests

Research on forest dynamics and water yield and quality

The relationship between the integrity of a catchment and its water yield was first explored in the late 1870s, with the idea that more trees would produce

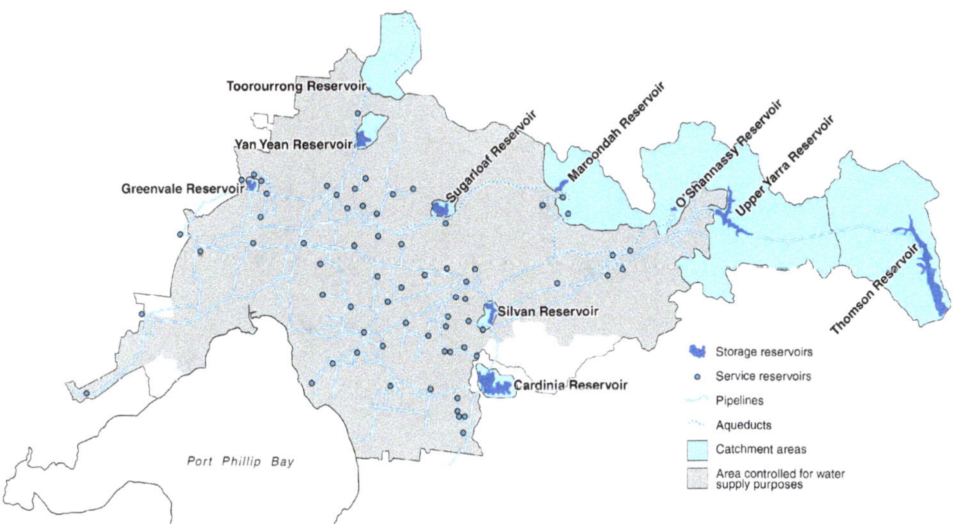

Figure 5.2: Melbourne water supply system in the late 1980s (redrawn from Dingle and Rasmussen[5]).

more rain. The decision to ban logging in water harvesting areas was based on this premise[3] and the closed catchment policy has resulted in the production of high-quality water and the conservation of some of the most ecologically intact forest in Australia.[11]

Understanding of the relationships between tree cover and local and regional precipitation has developed considerably since the 1870s. Research programs on Mountain Ash forests and water yield have been conducted since 1948.[68] The research on catchments has mainly focused on disturbance regimes affecting forests, such as fire, logging and road building, and how these regimes affect tree growth, water availability and water quality. For example, a long-term study commenced in the Maroondah Catchment in 1968 used clearfelling as a proxy for fire[68] but it also experimented with patch cutting and thinning.

Melbourne's closed water catchments have also been important in studying many aspects of the ecology and dynamics of Mountain Ash and Alpine Ash forest ecosystems. Work on forest and wildlife ecology commenced in the Upper Yarra, O'Shannassy and Maroondah catchments in mid-1983 and has continued ever since, making it among the longest-running work of its kind globally. There are ~80 long-term research sites permanently established within the three catchments and a further 80+

Figure 5.3: Thinning trees for Maroondah Catchment experiments in 1968 (image: Melbourne Water).

sites located outside the catchments. The 160+ sites occur across different land tenures, enabling scientists to study the differences between forests within catchments and forests outside the catchments, where human disturbances such as logging and farming occur. The sites are surveyed annually for plants, large old trees, arboreal marsupials, and birds. These sites are now formally part of Australia's Long-Term Ecological Research Network (http://tern.org.au/Long-Term-Ecological-Research-Network-pg17872.html).

Forest ecology research in the catchments has been led by a team at the Australian National University and resulted in a range of discoveries. These include: (1) old-growth Mountain Ash forests are the most carbon-dense forests on earth; (2) most of the carbon in a forest is not lost in a fire (including forests burned at very high severity) but remains part of regenerating forests; (3) the age of the forest has profound effects on the severity and spread of wildfires; and (4) some species of animals (e.g. the Yellow-bellied Glider, *Petaurus australis*) are strongly associated with old-growth Mountain Ash forest and are rare in or absent from younger forest.

Figure 5.4: The Mountain Ash is the tallest flowering plant in the world. The forests of the water catchment areas are spectacular, and support tall old-growth trees. This photo shows Site 470, one of the long-term research sites established by researchers from the Australian National University three decades ago (image: E. Beaton).

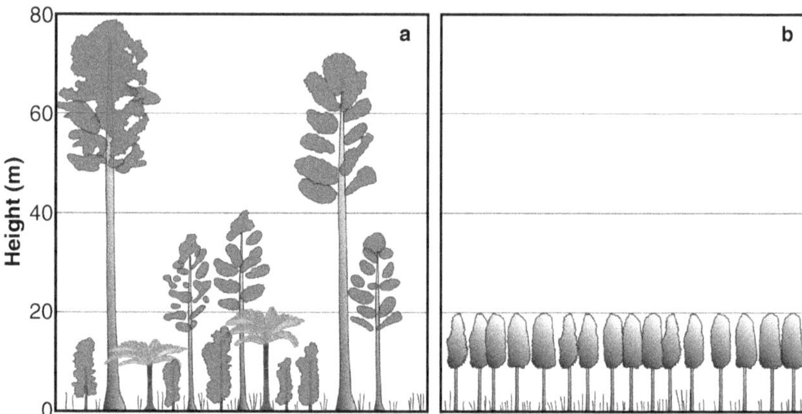

Figure 5.5: a) The complex structure of old forests. b) A single-aged stand of young trees (image: C. Hilliker).

Fire and water yield

Fire can have significant impacts on water yield and quality through its effects on the age and structure of the forest. High-severity fires in Mountain Ash and Alpine Ash forests can be stand-replacing events in which the original stands of trees are killed and replaced by a new cohort of regenerating stems. The loss of mature trees and the rapid regeneration of young trees in burnt areas increases evapotranspiration rates and reduces soil moisture. The evaporation of water from plant leaves draws moisture through stems up from the soil – a process called evapotranspiration. All trees require water as they transpire, but the rate of transpiration decreases with tree age. Young trees transpire at faster rates than older trees, which means more moisture is taken from the soil. Pulses of young regrowth trees lead to more rainfall being used by young trees rather than becoming runoff.[40] Another way that large old trees can increase water yield is through 'fog drip', which occurs when very tall trees capture water droplets from fog or very low cloud. The water droplets run down stems and branches and into the soil.[69]

Several studies have shown that fire and subsequent regeneration of Mountain Ash trees influence the amount of water flowing through the forest and into the reservoirs. In approximately the first five to 10 years following a fire, there is an increase in streamflow due to a decrease in the interception of rainfall from the canopy.[68] However, streamflows tend to decline in the 30–40 years following a large fire, as observed in the areas affected by the 1939 Black Friday fires.[40,68] It can take the streamflow up to 150 years after a major (stand-replacing) fire to recover to levels typical of ecologically mature Mountain Ash forest.[70] For example, studies in the

Maroondah Catchment found that old-growth Mountain Ash forests yield almost twice the amount of streamflow annually compared to regrowth forests comprising trees less than 25 years old.[71]

While Mountain Ash trees require fire to release and germinate seeds, immature trees cannot produce seeds until they are at least 25 years old. If Mountain Ash forests burn twice within 25 years, the forest will be replaced by stands of wattle (*Acacia* spp.) trees, delivering significantly reduced water yields.[72]

Fires can have other effects on water catchments. For example, fires burn the leaf litter and humus that protects the soil, making the soil vulnerable to erosion. Subsequent rain events in burnt areas result in large amounts of sediment being washed into watercourses.[39]

Logging and water yield

The effects of logging on water quality and water yield in the catchments have always been contentious. There have been arguments over land use every time a new water catchment has been proposed (see Chapters 2 and 3) but generally the catchment areas have been closed to logging activities. In the 1870s, supporters of the timber industry insisted that logging could occur in the Yan Yean Catchment without harming the water supply.[3] Similarly, in 1907, the Department of State Forests claimed that logging could occur harmoniously alongside water supply in the O'Shannassy Catchment.[5] These arguments were not supported by the MMBW or the state government, and we now know that disturbance of forests through logging has negative long-term effects on streamflow and water yield.[71] In addition, logging can increase sediment loads in water, via direct disturbance or erosion caused by road-building or vegetation removal.

Industrial logging occurs in the catchments of three tributaries of the Yarra River – Starvation, McMahons and Armstrongs creeks – along with substantial areas in the Thomson River Catchment where over 11 000 ha of Mountain Ash and Alpine Ash forest has been logged.[73,74] The forests of the Thomson River Catchment were available for logging for many decades before and after the development of the area as a water supply resource. The Yarra tributaries have always been subject to logging, even after they were diverted into the Upper Yarra system in the late 1960s. Under current prescriptions, forest on one of the Yarra tributaries is designated for logging in any given year.

Several studies in the ash-type eucalypt forests of the catchment have examined the effects of logging on water yield and quality. Felling of trees causes significant reduction in water yield, due to the high transpiration

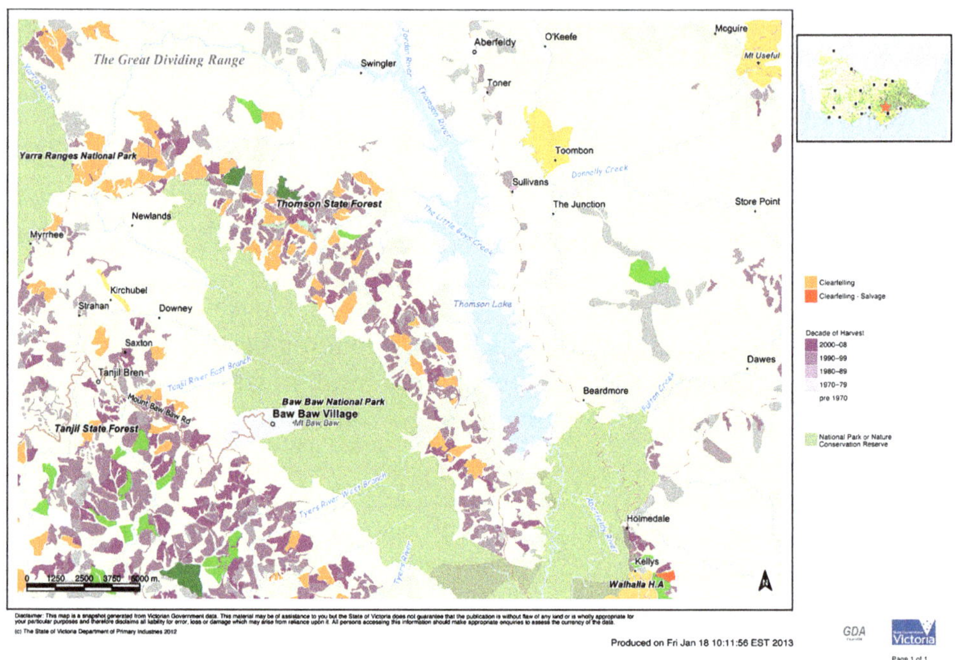

Figure 5.6: Map of the Thomson region, showing the history of logging in the catchment region (image reproduced from the Victorian Department of Sustainability and Environment).

rates of young regenerating trees, compared with old, established trees.[40,41,68,70,71] Logging also creates a young, dense regenerating forest that is fire-prone because of altered fuel characteristics[75] and altered micro-climate conditions (e.g. drying of the understorey forest via a reduction in canopy coverage of older trees).[72] Access roads for logging vehicles and the removal of trees contribute to soil erosion, which affects water quality.[76] There are turbidity sensors on the Yarra tributary pipelines to monitor water quality; supply can be halted if turbidity reaches high levels.

The end of the MMBW

Over time, the Greater Melbourne region (and the rest of Victoria) has been served by many water and sewerage authorities. In 1991 the state government instituted a series of major institutional reforms, and the Melbourne Water Corporation was formed[2] by amalgamation of the MMBW with other water boards located predominantly in south and south-eastern Melbourne (and surrounding areas), previously beyond the jurisdictional boundaries of the MMBW (e.g. Mornington Peninsula District, Emerald–Gembrook–Cockatoo Trust and Warburton Water Trust).[2] The new institutional

Box 5.2: Chlorination of Melbourne's water supply

Melbourne's water supply was considered 'pure' and did not require treatment because of the closed catchment policy. However, chlorination plants were installed at various locations after World War II and particularly during the 1960s and 1970s. These included plants upstream of Preston Reservoir on the Maroondah Aqueduct and the alienated (and logged) catchments such as those on the Yarra tributaries. However, it is likely that most of the chlorination plants contributed little to the protection of metropolitan water consumers as the plants were too remote to maintain residual chlorine levels in the water.

As the MMBW placed increasing emphasis on water quality, it adopted the National Health and Medical Research Council guidelines for water quality. A Water Quality Group was formed in Operations Division – for the first time the MMBW had personnel dedicated to the task of improving water quality and paying greater attention to the results of chemical and bacteriological water sampling. The adoption of the guidelines resulted in the installation of chlorination plants at the storage reservoirs that directly supplied the metropolitan area. Plants were located on the outlets of Yan Yean, Silvan, Greenvale and Cardinia reservoirs. As the main service reservoirs were open or uncovered, chlorination plants were also installed at some of those locations (e.g. Mount Waverley Reservoir). These service reservoirs have now been roofed or replaced with steel tanks.

arrangements shifted responsibility away from the experts in water supply (in the former MMBW) and handed it to a state government department responsible for water and environment. In essence, this meant that decision-making on infrastructure and water supply had to go through additional levels of bureaucracy and politicisation that had not previously been part of the MMBW's remit.

Corporatisation of water providers led to the contracting-out of design and construction work that had formerly been done by MMBW staff. This led to reduced levels of experience in both decision-making and the practical aspects of the former MMBW's activities. It also meant that many decisions were made on short-term political cycles of three to four years rather than over the long term, as had occurred previously.

In 1994, responsibility for water supply was partitioned among four bodies. Melbourne Water (the wholesalers) became responsible for catchment management, the harvest, storage and distribution of water to the major service reservoirs, associated infrastructure, and the wholesale supply of water. The retail supply of water to individual households and businesses was managed by retailers: Yarra Valley Water, South East Water and City West Water.[2] The partitioning of water supply infrastructure was achieved by designating 'interface points'. The retail companies controlled the assets involved with distributing water to consumers.

Catchment areas as national parks

Following the amalgamation of the water boards and the break-up of the MMBW's activities, the parks that had been managed by the MMBW became part of Parks Victoria. The MMBW had managed many metropolitan parks (e.g. Jells Park, Werribee Park and Westerfolds Park) in addition to the reservoir parks (e.g. the picnic areas at Maroondah and Cardinia reservoirs and Donnellys Creek). The reservoir parks were leased to Parks Victoria by Melbourne Water. However, Melbourne Water maintained access to the areas to service underground assets such as pipes, outlets and other infrastructure. At the same time, plans were being drawn up to gazette several catchment areas as national parks. Previously, the MMBW had resisted attempts to have its catchments converted to national parks, as it owned the land and was concerned about losing control of land management.

The Yarra Ranges National Park was gazetted in 1995, comprising the Maroondah, O'Shannassy and Upper Yarra catchments. Two key aspects of

Box 5.3: Acquisition of an open catchment – Tarago Reservoir

Following the amalgamation of the MMBW and other water boards in 1991, the newly formed Melbourne Water was given responsibility for wholesale supply of water. Melbourne Water took management responsibility for the Tarago Reservoir, north of Warragul. Tarago Reservoir was already supplying the towns of Warragul, Drouin and Neerim South and in the past had been a source of water for the Mornington Peninsula. When Cardinia Reservoir and the Thomson scheme were complete, supply of water from the Tarago Reservoir to the Mornington Peninsula ceased because of poor water quality.

Tarago is a multi-use (alienated) catchment, with highly productive farmland and logging areas that extended to the edges of the reservoir. Melbourne Water spent considerable time and effort developing strategies to increase the quality of water in the reservoir. Between 1994 and 2009, local farmers, business people and residents worked with Melbourne Water to improve management of the Tarago catchment. Tree-planting programs were implemented around streams and watercourses that fed into the reservoir. Farmers were given financial assistance to install crossing points over streams to prevent livestock walking through the water. Settling ponds were installed downstream of dairies and farms to arrest the flow of water into streams and the reservoir (F. Lawless, pers. comm.). A key element of the catchment management plan was that it would be ongoing to ensure the continued quality of the water. Following installation of a water treatment plant, Tarago Reservoir became part of water supply for the Mornington Peninsula in 2009.[86]

this national park are that it remains closed to the public and that Melbourne Water retains control over land use practices. Both decisions were fought for by Melbourne Water during the development of the terms of the management of the Yarra Ranges National Park (F. Lawless, pers. comm.).

Changes in water supply infrastructure

Concurrent with major institutional reforms, times also were changing for water supply infrastructure. For example, the O'Shannassy Aqueduct was falling into disrepair and was expensive to maintain. Its use was reduced in 1994 and it was completely decommissioned in 1998, after nearly 85 years of service. This resulted in a loss of ~30 GL of catchment yield annually, in addition to the loss of ~10 GL from Cement Creek which had been diverted into the O'Shannassy Aqueduct in the 1967–68 drought – a combined loss of harvestable yield from the O'Shannassy and Cement Creek catchments of 40 GL. This equates to ~10% of the annual consumption of the entire water supply system, a substantial proportion, particularly under drought conditions.

Prolonged drought: 1997–2009

The Thomson Reservoir first reached full capacity in 1990 and it became the largest single source of water for Melbourne. It filled again in 1996, enabling a significant reduction in the amount of water supplied from Sugarloaf Reservoir. This was important because water from Sugarloaf Reservoir is fully treated before distribution to the system, making it 'expensive water' compared with that from the Thomson Reservoir, which requires minimal treatment and is at an elevation which enables gravity-fed supply.

The Thomson Reservoir has a hydroelectric plant on its downstream outlet, which was used in the 1990s because the reservoir was full. Generating hydroelectricity was lucrative and raised substantial revenue via its sale into the power grid. The downside of generating hydroelectricity is that water running from the reservoir for this purpose is lost to the water supply system, but this is not a significant issue if streamflows are high and the Thomson Reservoir is close to capacity or overflowing.

The first prolonged drought after the completion of the Thomson Reservoir occurred from 1997 to 2009, the most severe drought in Victoria's history. It led to a significant draw-down of water from the Thomson Reservoir, as supplies also dwindled in other reservoirs. Reduced rainfall had serious implications for the Thomson Reservoir because it takes four years of average streamflow to fill this storage without the release of any water (see Chapter 4). Maintaining storage levels during drought conditions was therefore impossible.

The Sugarloaf Reservoir was returned to normal output in 1998 to counter dwindling supplies elsewhere. With the benefit of hindsight, it may have been more appropriate, and less expensive, to stop generating electricity and thereby reduce output from the Thomson Reservoir earlier and use more water from Sugarloaf Reservoir. However, in 1997 the focus was on generating revenue rather than on the integrity of the water supply to Melbourne. The decision to continue using Thomson Reservoir to generate electricity, while at the same time not utilising water from Sugarloaf Reservoir, added to water supply issues that were compounded by the unprecedented length of the drought.

New infrastructure projects

During the 1997–2009 drought, Melbourne's water storages dropped from 58% in January 2006 to 39% in December 2006, after the lowest annual

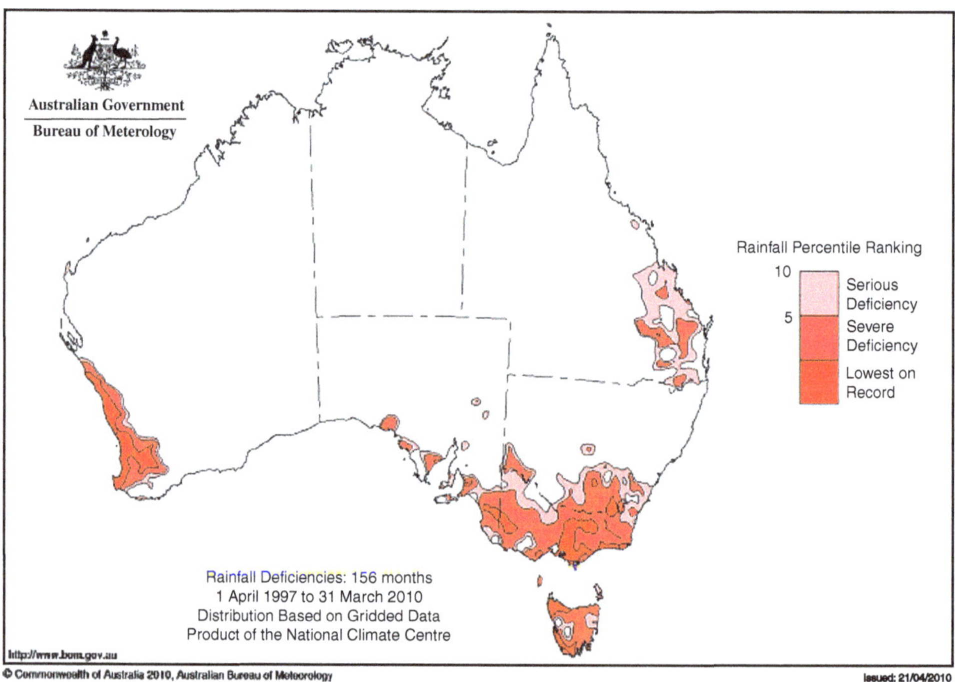

Figure 5.7: Rainfall percentiles (differences from the average conditions) for Australia, 1997–2010 (image: Bureau of Meteorology).

inflows on record.[77] By the middle of 2007, the major storage reservoirs were collectively at less than 30% capacity. They also dropped below 30% in mid-2009.[78]

After nearly a decade of drought, in June 2007 the then Premier, Steve Bracks, announced two new major water supply infrastructure projects – the construction of a pipeline from the Goulburn River downstream of Eildon Dam to Sugarloaf Reservoir in Melbourne (the North-South Pipeline) and a desalination plant at Wonthaggi, south-east of Melbourne.[79] These were the first major augmentations to the water supply system in 25 years and were intended to secure water for the continually expanding city for the next 25 years. Notably, both projects were a significant departure from the traditional approach of collecting water by building reservoirs and associated water distribution infrastructure.

The North-South Pipeline

The North-South Pipeline offtake is located on the Goulburn River downstream of Eildon Dam – the major water storage reservoir for irrigators in the Goulburn Valley. Construction of the 70 km pipeline between the Goulburn River and Sugarloaf Reservoir commenced in 2008 and was completed in early 2010 at a cost of $750 million (in 2010 values). Water flowed along the pipe between February and September 2010 but the pipeline has not been used since.[80]

Initially, the reason for halting the flow was that Sugarloaf Reservoir was full. In addition, there was adequate water available to Sugarloaf Reservoir from the Yarra River at Yering Gorge and from the Maroondah Reservoir. With a change of government in November 2010, the incoming Minister for Water, Peter Walsh, issued new rules for the use of the pipeline by amending the *Water Industry Act 1994*. These stated that if the total capacity of Melbourne's water storage was at 30% or less on 30 November of any given year, the pipeline could be brought into service.[81] However, the government added a let-out clause which allowed it to approve the operation of the pipeline under 'crisis' conditions.

Pros and cons

The location of the offtake for the North-South Pipeline from the Goulburn River downstream of the 3000 GL Eildon Dam allows water flows to be regulated and gives security in supplying water to meet demands from Melbourne via the pipeline. This is an advantage over other possible diversions such as those on the Big River or the Black River (see Chapter 4,

Box 5.4 and discussion below), which are upstream of Eildon Dam and subject to environmental fluctuations.

Various community groups strongly opposed the North-South Pipeline. Farmers from the Goulburn Valley were fiercely protective of 'their' water and were not impressed by the possibility of less water being available to them. The idea of 'their' water being used to flush Melbourne's toilets seemed to be a misuse of valuable irrigation water. The concept of water ownership is a moot point, however, as water is best supplied where it is most needed.

Another issue is that there is no need to use the pipeline during normal years, because there is more than sufficient water available from existing resources that supply Sugarloaf Reservoir. The current annual average yield from the Maroondah Catchment is 75 GL and this water is pumped into Sugarloaf Reservoir. In addition, in an average year, a further 200 GL could be pumped from the Yarra River at Yering Gorge into Sugarloaf Reservoir. The total of 275 GL represents two-thirds the volume of water consumed by Melbourne in any given year. However, the supply zone for the Sugarloaf Reservoir is limited and the downstream system is incapable of distributing such a large volume of water in a given year. Therefore, the allocation from the Goulburn River of up to 75 GL annually is unlikely to be required other than in extreme circumstances. This probably contributed to the government's decision to stop operation of the pipeline.

Desalination plant

Construction of the desalination plant began at Wonthaggi in 2009. Water from the plant first flowed in September 2012, almost a year later than planned, after long delays in construction and a cost overrun of more than $1 billion.[82] Water produced by the desalination plant is pumped from Wonthaggi to Cardinia Reservoir, from where it can be pumped to Silvan Reservoir for gravity-fed distribution to the parts of the system that Cardinia Reservoir cannot supply due to its relatively low elevation and geographic position.

To secure water supply using desalinated water, the Victorian government is required to place an order for water from the desalination plant's operators by April in a given year for the coming financial year. The minimum and maximum orders that can be placed are 50 GL and 150 GL, respectively, in 25 GL increments. For the 2012–13 financial year, the government made a request for zero GL of water due to the current levels in the major storage reservoirs. The government has opted to have the desalination plant shut down following its initial testing phase, until water from it is required.[83]

Pros and cons

The primary advantage of the desalination plant is that it is a significant departure from the traditional method of extracting water from surface resources through constructing dams and associated infrastructure. These traditional approaches are vulnerable to low water yields during droughts. Hence, the desalination plant provides Melbourne with a water supply independent of the impacts of climate change, drought, fire and their interaction.

Cost is a major factor in the use of a desalination plant. Payment for the plant is required to recover the capital costs of construction, even if no water is produced. Melburnians have to pay $650 million (in 2012 values) per year between 2013 and 2023,[84] ~60% of the total operating budget of Melbourne Water.[85] The annual cost rises by an additional $110 million if the full quota of 150 GL is supplied by the plant.[85]

There is unlikely to be a need for water from the desalination plant under normal weather and operating conditions at current annual rates of consumption, because of the architecture of the existing water supply system and the storages into which desalinated water might be pumped. Cardinia and Silvan reservoirs are the destinations for water flowing from other storage reservoirs and diversions upstream. These include the O'Shannassy, Upper Yarra and Thomson reservoirs and water from the Yarra tributaries and Coranderrk Creek. Desalinated water is likely to be required only if the system is under extreme stress.

If there is a need to produce desalinated water, the plant is entitled to supply the required amount at any time and rate of its choosing during a given financial year. It may seem intuitive to supply water at the times of the year when it will be in highest demand, but this is unlikely to occur because the logistics of running a desalination plant mean it may not be economically sound to supply water during the summer period. Electricity demands peak on very hot days, and it would not make economic sense to produce water during times of higher power tariff or to place additional stress on the electricity grid by having the desalination plant operating then. From an economic perspective, it would be in the interests of the desalination plant operators to provide water as a steady flow, using the equipment at low capacity over a long period to produce a given volume of water. Desalinated water will most likely be pumped to Cardinia Reservoir over a period of several months, and stored until needed. The level of water in Cardinia Reservoir will need to be constrained below full capacity to accommodate pre-ordered supplies of desalinated water.

Utilising the new infrastructure: the North-South Pipeline v. the desalination plant

With the establishment of the North-South Pipeline and the desalination plant, together with the protocols for their use, Melbourne Water is compelled to make detailed assessments of the status of the water supply system twice in a given year. This is done in April for the desalination plant (before the usual 'filling period' for the water storages) and at the end of November for the North-South Pipeline, in advance of the peak (summer) consumption period. There are key questions associated with the utilisation of the North-South Pipeline and the desalination plant: Which of the new water infrastructure alternatives should be used if Melbourne's water supply levels decline? Or should both be used?

All water is precious in a drought, regardless of where it comes from. While the pipeline offers flexibility for Melbourne's water supplies, it is still ultimately dependent on rainfall and subsequent inflows to the Goulburn River and Eildon Dam. The possible use of the pipeline would provide additional stability to Melbourne's water supply, but when a severe drought reduces inflows to Melbourne's water storage the weather conditions are also likely to affect catchments north of the Great Dividing Range.

Under drought conditions, the solution of choosing the cheapest option of water supply may be inappropriate. If prolonged drought occurs, it is likely that desalinated water, despite its higher cost, will be used to supply Melbourne in preference to water piped from the Goulburn River, as the rural communities do not have the benefit of an alternative supply. Moreover, under current legislation, the combined water storage of all major supply reservoirs must be at or below 30% on 30 November of any year for the pipeline to be used.[81] Water storage at 30% of total capacity is extremely low (virtually 'crisis' level). While Melbourne's storage did fall below 30% capacity three times during the 1997–2009 drought, it was *never* below that level in November of any year. Indeed, the total capacity of the water supply reservoirs has never been below the 30% level at the start of a summer except in 1984[80] when the newly completed but empty Thomson Reservoir distorted estimates of capacity levels. In that instance, the empty Thomson

Reservoir was counted in the overall tally for the water supply. The large size of the Thomson Reservoir resulted in a low value for the proportion of capacity, even though supplies were otherwise adequate at the time.

Large reductions in water supply storage over time are usually the result of significantly reduced inflow caused by drought. If water supply storage is declining, water from the desalination plant could be ordered in advance to ensure that levels do not fall drastically low. Desalinated water could be used to limit the rate of loss in reservoirs or increase the overall storage of reservoirs before levels drop to the trigger point of 30% capacity for the North-South Pipeline being brought into service.

The current decision to run the North-South Pipeline only in extreme circumstances, together with the availability of desalinated water, have effectively removed the pipeline as a practical supply option. On paper this seems problematic, especially when cost is taken into account. However, the option of using desalinated water is attractive as it is independent of the vagaries of weather and is therefore a guaranteed supply. Hence, the desalination plant could be viewed as a long-term strategy for providing water in times of dire need, not as a short-term solution to reduced inflows into storage reservoirs. The desalination plant may well become a critical part of the water supply system if projected population growth rates in Melbourne are maintained for several more decades.

Finally, options for water supply are more complex than we have suggested above. A carbon price on the water industry (e.g. through increasing the cost of power to run the desalination plant) could substantially increase the cost of water production using this method. This would apply not only to the desalination plant, which is very energy intensive, but also to the North-South Pipeline which also requires the operation of energy-intensive pumping stations and/or treatment plants to produce and deliver potable water. The desalination plant also involves other environmental impacts, such as the production of hyper-saline water, and the effects on the marine ecosystems into which it is discharged.

When construction of the North-South Pipeline was announced, the current state government was then in Opposition and at the time it deemed the project to be an expensive 'white elephant'. The policies of the current government enforce this 'white elephant' status, particularly with the desalination plant commissioned and ready for operation. It is essentially irrelevant whether the North-South Pipeline, the desalination plant or both is a superior option for augmenting Melbourne's water supply: the North-South Pipeline has effectively been decommissioned and no water is being ordered from the desalination plant. While recognising that both the North-South Pipeline and the desalination plant were proposed during a drought and it was not known how long the drought would persist, it appears that the two

Box 5.4: Did the government get drought management wrong?

The North-South Pipeline and the desalination plant projects were fiercely opposed. The Opposition complained that the combined cost of the projects, $4.9 billion, was far too mcuh. It also argued that the timing of the projects meant that Melbourne would not reap the benefits for many years.[79] However, we suggest that a broader and longer-term perspective about these projects is warranted.

By 1999, when the Labor government was elected and the drought had been going for two years, none of the augmentations to the water supply systems as suggested in various water supply strategies had been implemented. Yet the MMBW and Melbourne Water had stated repeatedly that augmentation was necessary by the late 1990s.[56,61] Therefore, in the early 2000s the water supply system was approaching its limits and was under significant stress.

In 2004, seven years after the commencement of the drought and after five years in power, the government released the 'Our Water Our Future' policy, which included, among other things, the decision that no more dams would be constructed.[87,88] The government had eliminated the traditional mechanism by which Melbourne's water supply had usually been augmented in the past. Perhaps it was a reasonable decision, as low rainfall does not fill a new reservoir any more than it does an existing one; water supply via reservoirs is wholly dependent on streamflow. For example, because the Big and Black rivers are upstream of Eildon Dam and therefore subject to environmental fluctuations, supply of water for the North-South Pipeline would be more secure than water from diversions on either of those watercourses. Moreover, diverting rivers such as the Big, Black, Aberfeldy, Yea and Macalister rivers (see Chapter 4) would have been pointless if there were no sustained rainfall in catchments. Given the length of the 1997–2009 drought, it may have been foolish to persist with strategies that 'sit back and wait for rain'. The government was proactive in this regard with its proposal for the desalination plant. While the plant may not be needed for many years, it does provide a source of water that is not reliant on rainfall and will be valuable in future droughts. The much-maligned North-South Pipeline from the Goulburn River is idle now, but it was proposed as an insurance policy against the continuing drought and was probably the only quick-fix supply option available.

Hindsight makes it easy to criticise government decisions. In fairness, however, the then government had inherited a water system that was reaching its limits for supply after being ignored for a decade, and was at the mercy of the longest drought in Melbourne's history. Greater attention to forward planning by successive governments may have led to a more thorough review of options and ultimately to better and possibly more cost-effective outcomes.

projects were a 'belt and braces' approach. That is, two projects were commissioned when one or the other would have been adequate.

Were there potential alternatives to the North-South Pipeline and the desalination plant?

In commissioning the North-South Pipeline and the desalination plant, the government had rejected or ignored other options that had been suggested by Ronalds in 1962 and in water supply strategies in the 1980s and 1990s (see Chapter 4).

The water supply strategies in 1962, 1981 and 1986 included a recommendation to divert the Big and Black rivers, which are tributaries of the Goulburn River upstream of Eildon Dam, into the Upper Yarra Reservoir and the Thomson Reservoir, respectively. The suggestions were rejected by the then Premier Henry Bolte and were not supported by the Parliamentary Public Works Committee in 1964 (see Chapter 4). However, Jim Viggers considered the Big and Black rivers to be potentially appropriate new water supply sources for Melbourne. Diversions on either or both of those rivers were feasible and had the advantage of being gravity-fed systems. Moreover, unlike the North-South Pipeline, both river systems were within existing catchment boundaries and would not require purchase of freehold land or relocation of businesses or farms. Jim Viggers strongly believed these options were superior to the planned North-South Pipeline. Water going into long-term storage at either Upper Yarra or Thomson reservoirs is at high elevations; the reservoirs can supply the whole system via gravity and water does not need to be pumped. By contrast, taking water from the Goulburn River downstream of the Eildon Dam required several pumping stations to move the water across the Great Diving Range and into Sugarloaf Reservoir. Jim Viggers felt so strongly about these issues that he wrote to the then Prime Minister, Kevin Rudd. His main concern was that all reasonable options had not been explored adequately and, while the North-South Pipeline would cost less money in the short term than a diversion on the Big or Black rivers, it would have higher ongoing operation and maintenance costs.

Another proposal was to divert the Aberfeldy River into the Thomson Reservoir. Jim Viggers strongly believed that the opportune time to build the diversion was when the Thomson Reservoir was being constructed – all the necessary manpower and resources were available and the project could be finished at a relatively low cost (see Boxes 4.12 and 4.13).

Although further augmentation in the form of extra dams and reservoir facilities could increase the capacity of Melbourne's water supply, potential infrastructure projects would be subject to the same limited supplies of water as the existing supply reservoirs during drought. It is therefore difficult to determine whether Melbourne would have fared better during the drought if the Aberfeldy, Macalister, Big, Black or other rivers had been part of the city's water supply. In addition, the cost of new infrastructure and its operation must be estimated and compared to the costs of water from the desalination plant. If the costs are found to be less than those of the desalination plant, they may be considered to be appropriate for development.

None of these river diversions or dams is likely to be considered in the short to medium term, because the desalination plant can augment supply. In addition, the environmental impacts of river diversions or dams would be subject to far greater scrutiny now than the level of environmental effects assessment that typically accompanied past major Victorian water infrastructure projects.

Conclusion

The years from 1985 to 2012 included times of contrast between optimism and despair for water supply. When the Thomson Reservoir filled to capacity twice during the 1990s, the knowledge that there was plenty of water may have given a false sense of security before the 12-year drought. The lack of augmentation of the water supply became evident as the drought continued, leading to two controversial projects – the North-South Pipeline from the Goulburn River to Melbourne, and the desalination plant. The overall management of water supply suffered upheaval with the abolition of the MMBW and its replacement by Melbourne Water and a suite of water retailers, which led to a new wholesale/retail arrangement for water supply.

Chapter 6: The future

Introduction

It is not possible to accurately predict the future of Melbourne's water supply. However, the lessons on supply, operations and development learned from the past can be used when examining future challenges. Many of the key issues facing Melbourne over the next 50–100 years will not be new. Droughts, Melbourne's expanding population and increasing physical size have been perennial problems. One issue, however, is relatively new – the threat to Melbourne's water supply from climate change – although its potential impacts were noted in successive water supply strategies from the 1980s.[56,60,61] How will climate change affect the water supply as it acts synergistically with other factors such as fire? In this chapter, we discuss several issues likely to face water suppliers in coming years. What strategies will ensure that enough water is available as Melbourne continues to grow? For how long will the current supply options be sufficient?

Water availability and distribution

The primary challenges associated with water supply encompass resource availability (i.e. less water during droughts) and distribution. The latter encompasses not only urban sprawl *per se*, but also the spatial separation between new areas of demand and the locations of the water catchments. Current growth areas are in the west and, to a lesser extent, the north of the city, while the main catchment areas are east of Melbourne. However, the ability to move water between reservoirs and service facilities ensures that water can be where it is needed. Indeed, different reservoirs are so interconnected to allow for water transfer that managing the supply is more flexible now than it was historically.

The ability to move water between reservoirs makes it possible to meet demands to maintain water storages at a required level, as agreed by Melbourne Water (which operates the supply system) and the water retailers (which sell water to consumers). An example is the Thomson system. Although the Thomson Reservoir is ~120 km east of Melbourne, its water can be transferred to Upper Yarra Reservoir from where it can go to Silvan Reservoir. From Silvan Reservoir, water can be supplied to Cardinia, Greenvale and Yan Yean reservoirs.

Demand for water

Melbourne's overall water requirements have been projected to increase, despite decreases in individual water usage, because the population will continue to grow. Demand for water has been projected to be 550 GL per year by 2055.[88] The maximum capacity of the combined water storages is 1812 GL and there is sufficient available water under average streamflow conditions to provide for a projected population of 5 million people by 2055. However, the Achilles heel of Melbourne's water supply system is that, like most cities worldwide, it is ultimately reliant on rainfall and subsequent streamflow. The construction of the desalination plant (see Chapter 5) was, in part, an effort to boost capacity and circumvent the problems of insufficient rainfall. Indeed, augmenting natural inflows of water from other sources provides a new level of flexibility for supply which is not possible when relying on rainfall alone.

Water supply strategies and options: the need for advance planning

Strategic water plans for Melbourne have been prepared at regular intervals, outlining the need for augmentation of the water supply and schemes for doing so. Between the completion of the Thomson Reservoir in 1983 and the commencement of the North-South Pipeline in 2009, no major water supply infrastructure projects were undertaken (see Chapter 5). A past challenge was to manage water supply and develop strategies during the 1997–2009 drought, when there was significant stress on the supply system. However, as Chapters 1–4 clearly showed, most of Melbourne's water supply solutions were planned decades in advance. This highlights the importance of identifying long-term solutions. We suggest that the rush to address water supply issues associated with the 1997–2009 drought led to expensive and seemingly inappropriate 'quick fixes', most notably the North-South Pipeline (see Chapter 5). The next major water supply plan is in preparation; the draft of the 2013 Melbourne Water Plan has been issued for public comment.[85]

Future threats to water supply

Three major factors have the potential to threaten the security of Melbourne's water supply – climate change, fire and logging. We discuss these separately below, but it is critical to recognise that they can interact strongly. For example, climate change can increase the prevalence of severe fires. Similarly, logging can make forests more fire-prone and at risk of burning at higher severity and over larger areas.[72,75]

Climate change

Climate change, coupled with other factors such as increasing population, will produce new challenges to the integrity of the water supply for Melbourne. Climate change projections indicate that Melbourne's average annual temperature will increase and average annual rainfall will be reduced significantly over the coming 40 years. The frequency of rainfall events and patterns of rainfall will also change, as will the number of days of extreme fire risk[67,89,90] (see Table 6.1).

Climate change is likely to reduce overall streamflow in catchments. For example, recurrent wildfires will reduce the age of the forest and increase the prevalence of young regrowth trees which transpire large amounts of water. Possible implications of climate change are greater dependence on water from the desalination plant and, to a lesser extent, the North-South Pipeline.

Fire

Fire can be a significant threat to water quality and yield. Many fires have occurred in the water catchments, the last major conflagration being the 2009 Black Saturday fires which burned ~93% of the O'Shannassy, 75% of the Maroondah and 2% of the Upper Yarra catchments. Smaller catchments, such as Wallaby and Armstrongs creeks, from which water flows into Yan Yean and Silvan reservoirs respectively, were completely burned. The total area burned was ~30% of the total catchment for Melbourne's water supply.[78]

The death of large trees following the 1939 and 1983 fires had long-term impacts on water yield and quality. Soil exposed by fire is susceptible to

Table 6.1. Climate change projections for Melbourne (data from CSIRO/BoM[90]).

Weather pattern	2020 projection	2050 projection
Average annual temperature	Increase by 0.3–1.0°C	Increase by 0.6–2.5°C
Average annual rainfall	Decrease by 5–0%	Decrease by -13–1%
Average no. hot days (>35°C)	Increase by 1–3	Increase by 3–9

erosion. Following the 2009 fires, workers at Melbourne Water installed silt fences and screen traps throughout the affected catchments and around water in-flow points to reduce the amount of sediment. Fire effects on catchment yields are also significant, due to post-fire regeneration of trees (see Chapter 5). Regeneration of forest after the 2009 fires has been rapid, with thousands of Mountain Ash seedlings observed per square metre in some places, especially where old-growth forest was burned. The large

Figure 6.1: The fire in the O'Shannassy Catchment was not uniform. Some parts burned, while other areas remained unscathed (image: H. Weaver).

numbers of young and rapidly growing plants in the catchments transpire most of the rainfall precipitated, leaving little water for runoff and streamflow. The effects of the 2009 fires on water yield are likely to be substantial, particularly given the large areas of fire-damaged forest in places such as the O'Shannassy Catchment, previously one of the highest-yielding catchments in the water supply system. Extensive stands of young regenerating forest in the O'Shannassy Catchment will lead to significantly reduced water yields over the next 150 years. Another fire within the next 20–30 years could eliminate stands of Mountain Ash and Alpine Ash,[72] leading to a major collapse in water yield.

The potential effects of climate change on fire regimes[91,92] and, in turn, patterns of forest cover mean that additional strategies may be warranted to limit the risks of the catchments being burned. A possible option is to create buffers comprising 1–2 km of unlogged forest adjacent to the boundaries of the water catchments, to reduce the chance of fire burning into a catchment from an adjacent area. Such buffer areas would be grown through to an ecologically mature stage in which stands of trees are less likely to burn at high severity than are young, logged stands.[72]

Logging and the value of water and trees

Logging operations reduce the age of a forest of trees to zero, with dense new stands of young regrowth trees regenerating after harvesting. Several studies have demonstrated that the highest yields of water are produced from old-growth forest and that logging operations have long-term negative effects on water yield (see Chapters 4 and 5). Pulpwood and timber harvesting is permitted in the Thomson and Yarra tributaries catchments. Logging-induced reductions in streamflows, particularly in high-yielding catchments such as the Armstrongs Creek Catchment, which have predominately south-facing topography, create the potential for significant conflict between the value of water and the value of timber in Victorian forests.

The logging industry in Victoria persists because of its perceived contribution to the state's economy. But what price do we put on forest values other than timber and pulpwood? Water values are rarely included in economic analyses of forests, and are often ignored in the development of logging plans. Key forest ecosystem services, such as the water yield, can be assigned dollar values. One study published in the mid-2000s on the Thomson Catchment suggested that the water was worth more than the timber.[73] In addition, logging in the Thomson Catchment can substantially reduce overall water yield through reducing the age of the forest.[93]

Studies on the comparative economic values of a forest suggest a need for very long intervals between logging operations at a site (>200 years, in some evaluations), or no logging at all.[94] Information on reduced water yields from logged catchments has particular importance in the context of the clear price of water, that resulted from the development of the desalination plant. Hence, the relative financial losses in water yield from one area (logged catchments) versus the costs of production in another (the desalination plant) are now transparent. We suggest that the cost of desalinated water would need to be less than the water forgone as a result of catchment logging for the public to receive maximum benefit from the latter use of public natural resources.

Logging and other forest values

A forest has intrinsic values beyond the water and timber it can produce. For example, trees store large amounts of carbon. Carbon storage is a key ecosystem service and is critical in attempts to reduce the amount of greenhouse gases in the atmosphere and to tackle climate change.[95] Old-growth Mountain Ash forests have the highest above-ground biomass carbon density in the world, up to 1867 t of biomass carbon per hectare.[11]

Large amounts of carbon are emitted during logging operations, including through the use of fossil fuel in road-building and harvesting, the decomposition of slash such as trees crowns and lateral branches left after logging, and post-fire regeneration burning. Studies of forestry operations in water catchments suggests that 286 t of carbon is emitted per hectare as a direct result of logging.[95] This is ~16 times more than the amount released by a fire, including an intense conflagration like the 2009 Black Saturday fires. With the advent of a price on carbon, comparisons of the relative values of carbon storage versus pulpwood and timber production indicate that leaving forests unlogged is the superior financial option.

A further issue encompasses the complex interrelationships between logging, fire, forest age and water yields. Several studies have demonstrated that logging increases the fire-proneness of Mountain Ash forests (see Chapter 5),[72] with corresponding negative impacts on water yield and carbon storage. On this basis, we suggest that the best management option is to leave forests in catchments unlogged to limit the risks of fire and maximise the chance of forests reaching an old-growth stage[72] when they will yield the greatest amount of water[71] and store the most carbon.[11,95]

Additional implications are the needs to protect water catchments from fire burning into them from surrounding wood production forests, and to retain the closed catchment policy.

There has been substantial overcommitment of forest to the pulpwood and timber industry for the past two or more decades, and supplies of sawlogs from wood production forests will be exhausted within the next 15 years.[96] This will lead to pressure from the pulpwood and timber industry to log water catchment forests. We argue that the water, carbon and other ecosystem values (e.g. for biodiversity) of the catchments must be maintained and pressure to revoke the closed catchment policy must be very strongly resisted.

Environmental impacts

Most Melburnians would be unaware of the extent to which river systems have been altered as a result of water supply infrastructure for water supply or irrigation purposes. There were no requirements to maintain environmental flows downstream of major works in the early development of Melbourne's water supply infrastructure and no emphasis by the community or other groups on the need for environmental flows. Indeed, few, if any, projects before those in the 1980s examined the wider environmental impacts of river diversions or dams on the integrity of aquatic biodiversity and aquatic ecosystems. An exception was the work on the environmental impacts of the Lower Yarra scheme near Yering Gorge and the Bend of Islands.[59] The paucity of consideration of environmental impacts is especially notable given the emphasis on closed catchment policies dating back more than a century and associated attempts to maintain the cover of old forest to maximise water yield and quality. There is no doubt that future water infrastructure proposals will involve substantial studies of their environmental impacts and the efficacy of strategies to mitigate those effects.

Melbourne Water is now required to maintain flows downstream of works at Cardinia, Upper Yarra and Thomson reservoirs and is allowed to pump from the Yarra River at Yering Gorge only when the flow is above specified levels. The requirements to maintain environmental flows started with the construction of Cardinia Reservoir in the 1970s. There are now substantial requirements to maintain environmental flows in the Yarra River at several locations between the Upper Yarra Reservoir and Yering Gorge.

The development of the water supply system

Melbourne's water supply has come a long way since the first carts transported water from The Falls at the Yarra River in the 1830s and 1840s (see Chapter 1). Expansion of both water supply infrastructure and the city's suburbs has provided constant challenges that require a detailed knowledge of the intricacies of the water supply system to make it run effectively.

Technology has improved the ability to manage the supply of water to Melbourne. Water storages do not empty overnight. Water levels vary seasonally and rise and fall slowly enough that they can be carefully monitored and regulated. Nevertheless, there is concern about the ability of technicians to deal with unexpected situations or problems with supply. Such concerns are not new, but problems are handled differently now from in the past through widespread automation of the system that regulates the water supply.

Technology has improved to the point that remote operation by smartphone is being implemented. Valves and conduits can be operated at the touch of a button as water flow and output are monitored by sensors that send information directly to an operator's phone. On a larger scale, the automated system allows operators to evaluate weather conditions and use computer programs to estimate levels of water use on a daily basis. The system links all the major supply and metropolitan service reservoirs, and will automatically adjust water flows at each site to meet expected consumption levels. A highly automated system can monitor usage carefully and offer more options to solve supply issues. As the sensitivity of predictive models develops further, movements of water can be tailored with greater precision, which will reduce stress on the water supply system. Increased efficiency of water usage will also reduce stress on the system, as per capita consumption is reduced.

In the 60 years before 1990, only three engineers held the job responsible for managing operation of the headworks and the regional distribution system. Jim Viggers was one of them. Jobs are no longer permanent and people no longer devote an entire career to managing the water supply system. There is a lack of knowledge transfer to newer staff within Melbourne Water, which means that the expertise of senior or retiring staff is not passed on. Therefore, detailed knowledge of the intricacies of the water supply system is dwindling. The story of the design of the spillway at the Upper Yarra Reservoir (Box 3.6), recounted by design engineer Frank Barnes, is a valuable example. Indeed, this book contains the only written record, as Barnes is the last man alive who worked on the design of the project.

There are other implications of increasing technology, increasing rates of staff turnover and the loss of corporate knowledge of the water supply system. Now, very few (if any) staff have both intimate and holistic knowledge of the water supply system. Such knowledge is pivotal in making well-informed decisions about the best options for developing the water supply infrastructure as well as moving water in the most efficient and cost-effective ways under varying circumstances. Indeed, we suggest that the paucity of intimate and holistic knowledge of the water supply system, coupled with healthy (over)doses of political interference, may have been important contributing factors in the expensive but seemingly inappropriate development of 'problem' infrastructure such as the North-South Pipeline, which is unlikely to be required other than in extreme circumstances (see Chapter 5).

Conclusion

This book has detailed the evolution of a great city's water supply. Central to this story is the role of the MMBW in developing the water supply system, primarily as an agency outside the direct influence of the state government. Some of the innovations, such as the closed catchment policy, were many decades ahead of their time and the development of the sciences of forest ecology and management and human epidemiology put the MMBW in the forefront of catchment management globally.

We can look back at visionaries like Blackburn, Davidson, Thwaites, Ritchie and, more recently, Ronalds, to gauge how their influence shaped the development of water supply for Melbourne. In 50–100 years, we wonder if the efforts of current decision-makers will be held in similarly high regard. This is not to say that the old ways were better. We live in different conditions now, and have access to technology for water production and conservation that was totally unknown to design engineers of the early 1900s. Ritchie ruffled feathers with his radical and divisive proposal to draw water from the Upper Yarra River (see Chapter 2), yet it has proved a lynchpin of Melbourne's water supply. Perhaps there will be parallel perspectives on the controversial desalination plant and how water is supplied to Melbourne in the years to come.

Most Melburnians give little thought to the water that flows from their taps, where it comes from and how it got to their homes. The city's abundant high-quality water results from the dedicated efforts of many people over many decades.

References

1. Gibbs GA (1915) *Water Supply Systems of the Melbourne and Metropolitan Board of Works*. DW Paterson Co., Melbourne.
2. CONTEXT (2010) *Regional Water Supply Heritage Study: Melbourne Region. Volume 1: Thematic Environmental History. Prepared for Melbourne Water*. CONTEXT, Melbourne.
3. Dingle T, Doyle H (2003) *Yan Yean: A History of Melbourne's Early Water Supply*. Public Record Office Victoria, Melbourne.
4. Davidson W (1920) Melbourne water supply. Read to the Royal Historical Society of Victoria, 29 June 1920. Royal Historical Society of Victoria, Melbourne.
5. Dingle T, Rasmussen C (1991) *Vital Connections: Melbourne and its Board of Works 1891–1991*. Penguin, Melbourne.
6. Jackson MB (1859) On the water supply to the city of Melbourne, South Australia: comprising a brief description of the Melbourne gravitation water-works. *Minutes of the Proceedings of the Institution of Civil Engineers* **18**, 363–384.
7. Ritchie EG (1934) Melbourne's water supply undertaking. *Journal of the Institution of Engineers, Australia* **6**, 379–382.
8. MMBW (1908) *Photographic Views*. Osboldstone & Co., Melbourne.
9. Seeger RC (1960) *The History and Development of Melbourne's Water Catchment Policy*. MMBW, Melbourne.
10. Melbourne Water (n.d.) History of managing drought in Melbourne. Melbourne Water, Melbourne.
11. Keith H, Mackey BG, Lindenmayer DB (2009) Re-evaluation of forest biomass carbon stocks and lessons from the world's most carbon-dense forests. *Proceedings of the National Academy of Sciences of the United States of America* **106**(28), 11635–11640. doi: 10.1073/pnas.0901970106
12. McNicoll R (1981) Davidson, William (1844–1920). *Australian Dictionary of Biography*, The Australian National University. At http://adb.anu.edu.au/biography/davidson-william-5905/text10057. (Accessed 29 August 2012.)
13. LaNauze R (2011) William Thwaites: the engineer who changed Melbourne. At http://heritageaustralia.com.au/magazine.php?article=423. (Accessed 23 August 2012.)
14. Allen HB, Akehurst AP, Masson O, Hodgkinson C, McCrea W, Girdlestone TM (1889) *Royal Commission to Inquire into and Report upon the Sanitary Conditions of Melbourne. Progress Report Parts 1 and 2*. Government Printer, Parliament of Victoria, Melbourne.
15. Steere-Williams J (2010) Milk-borne typhoid and epidemiological practice in late Victorian Britain. *Journal of the History of Medicine and Allied Sciences* **65**(4), 514–545. doi:10.1093/jhmas/jrq010.

16. Smith FB (2002) Disputes about typhoid fever in Victoria in the 1870s. *Health and History* **4**, 1–18. doi:10.2307/40111435
17. Allen HB, Akehurst AP, Masson O, Hodgkinson C, McCrea W, Girdlestone TM (1889) *Second Progress Report of Royal Commission to Inquire into and Report upon the Sanitary Conditions of Melbourne.* Government Printer, Parliament of Victoria, Melbourne.
18. Preston H (1966) Blackburn, James (1803–1854). *Australian Dictionary of Biography*, Australian National University. At http://adb.anu.edu.au/biography/blackburn-james-1789/text2019. (Accessed 30 August 2012.)
19. Edwards D (1978) *Yan Yean: A History.* Yan Yean School Council, Melbourne.
20. Jessop JC (1942) *A Historical Survey of Melbourne's Water Supply.* MMBW, Melbourne.
21. Binnie GM (1981) *Early Victorian Water Engineers.* Thomas Telford, London.
22. Postel SL, Thompson BH (2005) Watershed protection: capturing the benefits of nature's water supply services. *Natural Resources Forum* **29**, 98–108. doi:10.1111/j.1477-8947.2005.00119.x.
23. Dingle T (1990) Thwaites, William (1853–1907). *Australian Dictionary of Biography*, Australian National University. At http://adb.anu.edu.au/biography/thwaites-william-8811/text15455. (Accessed 30 August 2012.)
24. MMBW (1900) *The Melbourne and Metropolitan Board of Works Sewerage Scheme: A History of its Inception and a Description of the Engineering Works Carried on in Connection with the Scheme.* Periodicals Publishing Company, Melbourne.
25. Halliday S (2001) Death and miasma in Victorian London: an obstinate belief. *British Medical Journal* **323**, 1469–1471. doi:10.1136/bmj.323.7327.1469.
26. Thwaites W (1902) *Engineer in Chief's Report 1901–1902.* MMBW, Melbourne.
27. McDonald P (2010) Melbourne: demography. In *eMelbourne: The City Past and Present.* School of Historical Studies, University of Melbourne. http://www.emelbourne.net.au/biogs/EM00455b.htm.
28. Melbourne Water (2012) *Annual Report 2011–12.* Melbourne Water, Melbourne.
29. Davis J, Murray S, Burchell FN (1902) *Interstate Royal Commission on the River Murray, representing the States of New South Wales, Victoria, and South Australia: Report of the Commissioners. With Minutes of Evidence, Appendices, and Plans.* Sands & McDougall, Melbourne.
30. Leonard EA (1909) *Melbourne Water Supply.* MMBW, Melbourne.
31. Australian Wood Pipe Company (1925) *Pioneer Wood Pipe.* Arthur Smyth & Sons, Sydney.
32. Gibbs GA (1925) *Water Supply and Sewerage Systems of the Melbourne Metropolitan Board of Works.* Engineering Publishing Co., Melbourne.
33. Sadlier F (2003) *Along the Length: An Account of Living and Working on the Maroondah Aqueduct.* Yarra Glen & District Historical Society, Yarra Glen.

34. MMBW (1956) *100 Years of Water Supply: A Chronological Summary.* MMBW, Melbourne.
35. Anon. (14 May 1927) 'Maroondah Reservoir'. *Healesville and Yarra Glen Guardian.*
36. Kelso A (1934–35) The construction of Silvan Dam, Melbourne water supply. *Proceedings of the Institution of Civil Engineers* **239**, 403–446. doi:10.1680/imotp.1935.15284
37. Ritchie EG (1922) *Future Water Supply of Melbourne.* MMBW, Melbourne.
38. Ritchie EG (1925) Water conservation. In *Save Australia: A Plea for the Right Use of our Flora and Fauna.* (Ed. J Barrett) pp. 159–173. Macmillan & Co., Melbourne.
39. Ritchie EG (1923) Forests in relation to water conservation. *Australian Forestry Journal* **6**(1), 89–95.
40. Langford KJ (1976) Change of yield of water following a bushfire in a forest of *Eucalyptus regnans. Journal of Hydrology* **29**, 87–114. doi:10.1016/0022-1694(76)90007-X.
41. Cornish PM, Vertessy R (2001) Forest age-induced changes in evapotranspiration and water yield in a eucalypt forest. *Journal of Hydrology* **242**, 43–63. doi:10.1016/S0022-1694(00)00384-X.
42. Department of Planning and Com munity Development (2012) Melbourne's water usage. At http://www.dpcd.vic.gov.au/_data/assets/pdf_file/0020/31367/water_consumption_1.pdf. (Accessed 1 October 2012.)
43. Noble WS (1977) *Ordeal by Fire: The Week a State Burned Up.* Hawthorn Press, Melbourne.
44. Jarrett PH, Petrie AH (1929) The vegetation of the Black Spur region: a study of some *Eucalyptus* forests. II. Pyric succession. *Journal of Ecology* **17**, 249–281. doi:10.2307/2256043.
45. Morgan M (1988) Ritchie, Edgar Gowar (1871–1956). *Australian Dictionary of Biography*, Australian National University. At http://adb.anu.edu.au/biography/ritchie-edgar-gowar-8215/text14375. (Accessed 30 August 2012.)
46. Foley JC (1957) *Droughts in Australia: Review of Records from Earliest Years of Settlement to 1955. Bulletin 43.* Bureau of Meteorology, Melbourne.
47. Stretton LEB (1939) *Report of the Royal Commission to Inquire into the Causes of and Measures Taken to Prevent the Bush Fires of January 1939, and to Protect Life and Property. And, the Measures to be Taken to Prevent Bush Fires in Victoria and to Protect Life and Property in the Event of Future Bush Fires.* Government Printer, Melbourne.
48. Olsen SJ, Bleasdale SC, Magnano AR, Landrigan C, Holland BH, Tauxe RV, Mintz ED, Luby S (2003) Outbreaks of typhoid fever in the United States, 1960–99. *Epidemiology and Infection* **130**, 13–21. doi:10.1017/S0950268802007598.
49. CONTEXT (2007) *Victorian Water Supply Heritage Study. Volume 1: Thematic Environmental History. Prepared for Heritage Victoria.* CONTEXT, Melbourne.
50. Anon. (10 April 1940) 'Upper Yarra Reservoir £2,000,000 project'. *The Argus.*

51. Tunaley SM (2007) *Upper Yarra Dam: A 1950s Construction Town.* Self-published by author, Yarrambat.
52. Ronalds AF (1962) *Future Water Supply of the Melbourne Metropolitan Area.* MMBW, Melbourne.
53. MMBW (n.d.) Thomson River development. (pamphlet). MMBW, Melbourne.
54. Campbell WM, Floyd WL, Trewin TC, Wilton J, Scanlan AH (1967) *Final Report on the Melbourne Metropolitan Future Water Supply Inquiry.* Parliament of Victoria, Melbourne.
55. MMBW (1975) *Report on Environmental Study into Thomson Dam and Associated Works. Volume 1: Proposed Works and Existing Environment.* MMBW, Melbourne.
56. Chapman M, Grove L, Paynter R (1991) *A Review of Future Water Supply Sources for Melbourne.* Melbourne Water, Melbourne.
57. Melbourne Water (2012) Water conservation. At http://www.melbournewater.com.au/content/water_conservation/drought_management/history_of_managing_drought.asp. (Accessed 18 October 2012.)
58. Thiess Contractors (1984) *Thomson Dam Concrete Structures and Associated Works.* Diamond Press, Brisbane.
59. MMBW (1974) *Lower Yarra Water Supply Development: Report on Yarra Brae–Sugarloaf Environmental Study.* MMBW, Melbourne.
60. Langford KJ, Hortolanyi IS, Molloy JD, Peterson KN (1982) *A Water Supply Strategy for Melbourne: Planning for an Uncertain Future.* MMBW-W-0075. MMBW, Melbourne.
61. Jackson G, Guttmann P, McIllree B (1986) *A Water Plan for a Growing Melbourne: Water Supply Strategy Review.* MMBW, Melbourne.
62. Department of Sustainability and Environment (2007) *Our Water Our Future: The Next Stage of the Government's Water Plan.* Victorian Government, Melbourne.
63. Anon. (April 1983) 'The day Victoria caught fire'. *Mokera* (MMBW staff newspaper).
64. National Ready Mixed Concrete Association (2012) Testing the compressive strength of concrete. At http://www.nrmca.org/aboutconcrete/cips/35p.pdf. (Accessed 12 December 2012.)
65. MMBW (1980) *The future of Rawson, Thomson Dam Works Township. Final Report of the Study Group.* Town and Country Planning Board, Shire of Narracan.
66. ABS (2011) Census data: Rawson, Victoria. Australian Bureau of Statistics, Canberra. At http://www.censusdata.abs.gov.au/census_services/getproduct/census/2011/quickstat/SSC21128?opendocument&navpos=95. (Accessed 26 October 2012.)
67. Howe C, Jones RN, Maheepala S, Rhodes B (2005) *Implications of Potential Climate Change for Melbourne's Water Resources.* CSIRO, Melbourne.
68. Jayasuriya MDA, Dunn G, Benyon R, O'Shaughnessy P (1993) Some factors affecting water yield from mountain ash (*Eucalyptus regnans*) dominated forests in south-east Australia.

Journal of Hydrology **150**, 345–367. doi:10.1016/0022-1694(93)90116-Q.

69. O'Connell MJ, O'Shaughnessy P (1975) *The Wallaby Creek Fog Drip Study*. MMBW-W-0004. MMBW, Melbourne.
70. Vertessy R, Watson F, O' Sullivan S, Davis S, Campbell R, Benyon R, Haydon S (1998) *Predicting Water Yield from Mountain Ash Forest Catchments*. CRC for Catchment Hydrology, Industry Report 98/4.
71. Vertessy R, Watson F, O'Sullivan S (2001) Factors determining relations between stand age and catchment water balance in mountain ash forests. *Forest Ecology and Management* **143**, 13–26. doi:10.1016/S0378-1127(00)00501-6.
72. Lindenmayer DB, Hobbs RJ, Likens GE, Krebs CJ, Banks SC (2011) Newly discovered landscape traps produce regime shifts in wet forests. *Proceedings of the National Academy of Sciences of the United States of America* **108**(38), 15887–15891. doi:10.1073/pnas.1110245108.
73. Alaouze CM (2004) The effect of conservation value on optimal forest rotation. *Land Economics* **80**, 209–223. doi:10.2307/3654739.
74. VicForests (2012) Timber Release Plan 2011–16. At http://www.vicforests.com.au/timber-release-plans.htm. (Accessed 22 January 2013.)
75. Lindenmayer DB, Hunter ML, Burton PJ, Gibbons P (2009) Effects of logging on fire regimes in moist forests. *Conservation Letters* **2**, 271–277. doi:10.1111/j.1755-263X.2009.00080.x.
76. MMBW (1980) *Catchment Hydrology Research: Summary of Technical Conclusions to 1979*. MMBW, Melbourne.
77. Melbourne Water (2012) Water storage history. At http://www.melbournewater.com.au/content/about_us/history_and_heritage/our_history_-_a_timeline.asp. (Accessed 22 November 2012.)
78. Melbourne Water (2010) *Annual Report 2009–10*. Melbourne Water, Melbourne.
79. Rood D, Guerrera O, Kleinman R (2007) 'Bracks' $4.9bn water plan'. *The Age*, 20 June 2007. http://www.theage.com.au/news/national/water-bills-set-to-double/2007/06/19/1182019116320.html.
80. Arup T (2011) '$750 million project down the drain'. *The Age*, 15 November 2011. http://www.theage.com.au/victoria/750-million-project-down-the-drain-20111114-1nfmz.html.
81. Victorian Government (2012) *Special Gazette* no. S108, 29 March. Victorian Government, Melbourne.
82. Schneiders B (2011) 'A year late and a financial disaster: desal companies come clean'. *The Age*, 28 October 2011. http://www.theage.com.au/victoria/a-year-late-and-a-financial-disaster-desal-companies-come-clean-20111027-1mm7k.html.
83. Schneiders B (2012) 'Desal plant reaches target – one year late'. *The Age*, 9 October 2012. http://www.theage.com.au/victoria/desal-plant-hits-target--a-year-late-20121009-27bay.html.
84. Wells R (2012) 'Desal water runs but no drinking for years'. *The Age*, 27

September 2012. http://www.theage.com.au/victoria/desal-water-runs-but-no-drinking-for-years-20120926-26lnq.html.
85. Melbourne Water (2012) *Water Plan 2013 (Draft).* Melbourne Water, Melbourne.
86. Melbourne Water (2012) Tarago Reservoir. At http://www.melbournewater.com.au/content/current_projects/water_supply/tarago_reservoir_upgrade/tarago_reservoir_upgrade.asp. (Accessed 16 December 2012.)
87. Moran A (2008) *Water Supply Options for Melbourne.* Institute of Public Affairs, Melbourne.
88. Lovering J (2005) *Water Supply–Demand Strategy for Melbourne 2006–2055.* Water Smart, Melbourne.
89. Clarke H, Lucas C, Smith P (2013) Changes in Australian fire weather between 1973 and 2010. *International Journal of Climatology* **33**, 931–944. doi:10.1002/joc.3480.
90. CSIRO/BoM (2007) *Climate change in Australia.* CSIRO and Bureau of Meteorology, Melbourne.
91. Cary GJ, Bradstock RA, Gill AM, Williams RJ (2012) Global change and fire regimes in Australia. In *Flammable Australia: Fire Regimes, Biodiversity and Ecosystems in a Changing World.* (Eds RA Bradstock, AM Gill and RJ Williams) pp. 149–169. CSIRO Publishing, Melbourne.
92. Williams RJ, Bradstock RA, Cary GJ, Enright NJ, Gill AM, Liedloff AC, Lucas C, Whelan RJ, Andersen AN, Bowman DJ, Clarke PJ, Cook GD, Hennessy KJ, York A (2009) *Interactions between Climate Change, Fire Regimes and Biodiversity in Australia. A Preliminary Assessment.* Department of Climate Change and Department of the Environment, Water, Heritage and the Arts, Canberra.
93. Peel M, Watson F, Vertessy R, Lau A, Watson I, Sutton M, Rhodes B (2000) *Predicting Water Yield Impacts of Forest Disturbance in the Maroondah and Thomson Catchments using the Macaque Model.* CRC for Catchment Hydrology, Melbourne.
94. Clarke HR (1994) Forest rotation and streamflow benefits. *Australian Forestry* **57**, 37–44. doi:10.1080/00049158.1994.10676111
95. Keith H, Lindenmayer DB, Mackey BG, Blair D, Carter L, McBurney L (2012) *Carbon Stocks and Impacts of Disturbance in Native Eucalypt Forest Ecosystems in the Central Highlands Catchments Supplying Water to Melbourne. Report to Melbourne Water.* Australian National University, Canberra.
96. Lindenmayer DB, Blanchard W, McBurney L, Blair D, Banks S, Likens GE, Franklin JF, Stein J, Gibbons P (2012) Interacting factors driving a major loss of large trees with cavities in an iconic forest ecosystem. *PLoS ONE* **7**, e41864. doi:10.1371/journal.pone.0041864.

Index

www.ingramcontent.com/pod-product-compliance
Lightning Source LLC
LaVergne TN
LVHW081300100826
845148LV00005B/935

* 9 7 8 1 4 8 6 3 0 0 0 6 8 *